The R.A.M.S. Library of Alchemy

Volume 32

Four Works of Roger Bacon
by

Roger Bacon

Radix Mundi

The Mirrour of Alchimy

The Oil of Antimony

Miracles of Art, Nature and Magick

R.A.M.S. Publishing Company

Four Works of Roger Bacon
by
Roger Bacon

Radix Mundi

The Mirrour of Alchimy

The Oil of Antimony

Miracles of Art, Nature and Magick

Produced by

Restorers of Alchemical Manuscripts Society

R.A.M.S. Publishing Company

R.A.M.S. Publishing Company
117 Rutherford Lane
Stuarts Draft VA 24477

Four Works of Roger Bacon

First Edition 2015

ISBN-13 **978-1511758253**
ISBN-10 **1511758252**

Image Processing by Philip N. Wheeler

Printed in the United States of America

Table of Contents

Dedicated to Hans W. Nintzel,

American Alchemist

and

Founder of the

Restorers of Alchemical Manuscripts Society

(R.A.M.S.)

Disclaimer

Liability: The publisher does not warrant or assume any legal liability or responsibility for the accuracy, completeness, or usefulness of any information, apparatus, product, or process disclosed. The publisher makes no representation as to the accuracy or completeness of the contents of this book and specifically disclaims any implied warranty of merchantability or fitness for a particular purpose. No warranty may be created or extended by written sales materials or sales representatives. You should obtain professional consultation where appropriate. The publisher shall not be liable for any loss of profit or other commercial or personal damages, including but not limited to special, incidental, consequential, or other damages.

Roger Bacon

Introduction

Philip N. Wheeler

Roger Bacon ((1210 to 1215?)-1294) was an English Alchemist and Philosopher during the Middle Ages who insisted on conducting his own experiments and observing the results, as opposed to depending upon the writings of others.

Hans Nintzel selected these four works for inclusion in the R.A.M.S. Library:

Radix Mundi

The Mirrour of Alchimy

The Oil of Antimony

Miracles of Art, Nature and Magick

Roger Bacon the Englishman.

RADIX MUNDI

By Roger Bacon

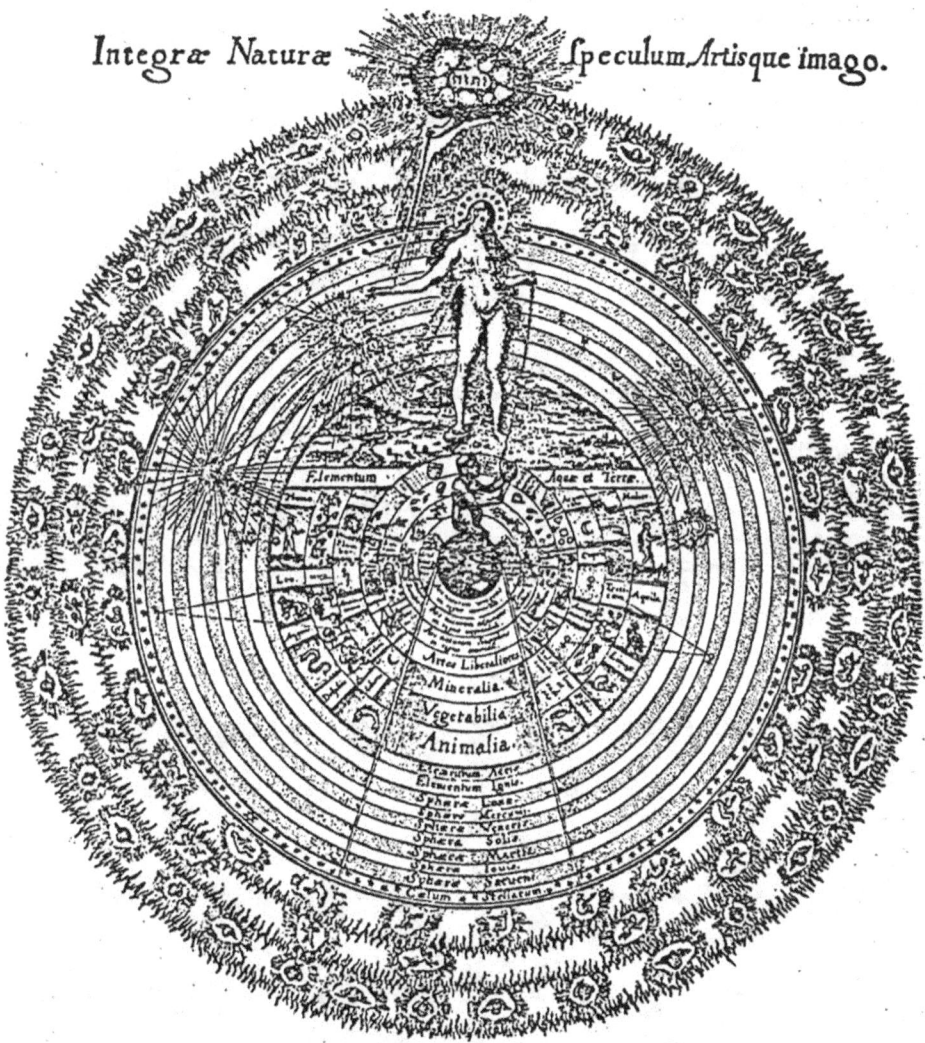

Integræ Naturæ Speculum Artisque imago.

Translated by William Salmon,
Professor of Physick

CHAPTER XXXVII

OF THE ORIGINAL OF METALS, AND PRINCIPLES OF THE MINERAL WORK

I. The Bodies of all Natural Things being as well perfect as imperfect from the Original of time, and compounded of a quaternity of Elements or Natures, viz., Fire, Air, Earth, Water, are conjoyned by God Almighty in a perfect Unity.

II. In these four Elements is hid the Secret of Philosophers: The Earth and Water give Corporeity and Visibility: The Fire and Air, the Spirit and Invisible Power, which cannot be seen or touched but in the other two.

III. When these four Elements are conjoyned, and made to exist in one, they become another thing; whence it is evident, that all things in Nature are composed of the said Elements, being altered and changed.

IV. So Saith *Rhasis, Simple Generation and Natural Transformation in the Operation of the Elements.*

V. But it is necessary that the Elements be of one kind, and not divers, to wit, Simple: For otherwise neither Action nor Passion could happen between them: So Saith *Aristotle, There is no true Generation, but of things* agreeing *in Nature.* So that things be not made but according to their Natures.

VI. The Eldar or Oak Trees will not bring forth Pears; nor can you gather Grapes of Thorns, or Figs of Thistles, things bring not forth, but only their like, or what agrees with them in Nature, each Tree its own Fruit.

VII. Our Secret therefore is to be drawn only out of those things in which it is. You cannot extract it out of Stones or Salt, or other Heterogene Bodies: Neither Salt nor Alum enters into our mystery: But as *Theophrastus* saith, *The Philosophers disguise with Salts and Alums,* the *Places of the Elements.*

VIII. If you prudently desire to make our Elixir, you must extract it from a Mineral Root: For as *Geber* saith, *You must obtain the perfection of the Matter from the Seeds thereof.*

IX. Sulphur and Mercury are the Mineral Roots, and Natural Principles, upon which Nature herself acts and works in the Mines and Caverns of the Earth, which are Viscous Water, and Subtil Spirit running through the Pores, Veins, and Bowels of the Mountains.

X. Of them is produced a Vapour or Cloud, which is the substance and body of Metals united, ascending, and reverberating upon its own proper Earth, (As Geber sheweth) even till by a temperate digestion through the space of a Thousand Years, the matter is fixed, and converted into a Mineral Stone, of which metals are made.

XI. In the same manner of Sol which is our Sulphur, being reduced into Mercury by Mercury, which is the Viscous Water made thick, and mixt with its proper Earth, by a temperate decoction and digestion, ariseth the Vapour or Cloud, agreeing in nature and substance with that in the Bowels of the Earth.

XII. This afterwards is turned into most subtil water, which is called the Soul, Spirit, and Tincture, as we shall hereafter shew.

XIII. When this Water is returned into the Earth, (out of which it was drawn) and every way spreads through or is mixed with it, as its proper Womb, it becomes fixed. Thus the Wise man does that by Art in a short time, which Nature cannot perform in less than the Revolution of a Thousand Years.

XIV. Yet notwithstanding, it is not We that make the metal, but Nature herself that does it. Nor do or can we change one thing into another; but it is Nature that changes them. We are no more than meer Servants in the work.

XV. Therefore *Nedus* in *Turba* Philosophorum, saith, Our Stone *naturally contains in it* the *Whole Tincture.* It is perfectly made in the Mountains and Body of the Earth; yet of itself (without art) it has no life or power whereby to move the Elements.

XVI. Chuse then the natural Minerals, to which, by the advice of Aristotle, add Art: For Nature generates Metaline Bodies of the Vapours, Clouds, or Fumes of Sulphur and Mercury, to which all the Philosophers agree. Know therefore the Principles upon which Art works, to wit, the Principles or beginnings of Metals: For he that knows not these things shall never attain to the perfection of the Work.

XVII. *Geber* saith, *He who has not in himself* the *knowledge, of* the Natural *Principles, is far from attaining the perfection of the* Art: being Ignorant of the Mineral Root upon which he should work.

XVIII. *Geber* also farther saith, That *our Art is only* to *be understood and Learned through the true wisdom and knowledge of Natural things:* that is, with a wisdom searching into the Roots and Natural principles of the matter.

XIX. Yet saith he, my Son, I shew thee a Secret, through thou knowest the Principles, yet therein thou canst not follow Nature in all things. Herein some have erred, in Essaying to following Nature in all her properties and differences.

The Ultimate Goal

CHAPTER XXXVIII

OF THE MERCURY, THE SECOND PRINCIPLE OF THE WORK.

I. The second Principle of our Stone is called *Mercury,* which some Philosophers call (as it is simple of itself) a Stone. One of them said, *This is a Stone, and no Stone, and that without which Nature never performs anything; which enters into, or is swallowed up of other Bodies, and also swallows them up.*

II. This is simply ARGENT VIVE, which contains the Essential Power, which Explicates the Tincture of our Elixir or Philosophers Stone.

III. Therefore saith *Rhasis, Such* a *thing may be made of it which exceedeth the highest perfection of Nature.* For it is the Root of Metals, Harmonises with them, and is the Medium that explicates and conjoyns the Tinctures.

IV. For it swallows up that which is of its own Nature and production; but rejects what is Forreign and Heterogene: being of an Uniform substance in all its parts.

V. Wherefore our Stone is called Natural, or

Mineral, Vegetable, and Animal, for it is Generated in the Mines, and is the Mother or Womb of all Metals, and by projection coverts into Metals: it Springs or Grows like a Vegetable: and abounds with Life like an Animal, by peircing with its Tincture, like Spirit and Life, everywhere, and through all particles.

VI. *Morien* saith, *This Stone is no Stone that can Generate a living Creature.* Another saith, *It is cast out upon the Dunghill as a vile thing, and it is hidden from the Eyes or understandings of Ignorant Men.*

VII. Also in *Libro Speculi Alchymiae,* it is said, Our Stone is a thing rejected, but found in Dunghills (i.e. in putrefaction, or the Matter being putrefied) containing in itself the four Elements, over which it Triumphs, and is certainly to be perfected by humane industry.

VIII. Some make MERCURY OF LEAD, Thus: Rx *Lead, melt it six or Seven times, and quench it in Salt* Armoniac *dissolved, of which take* lb. iij Sal

Vitrioli, lb j. Borax lb. $\mathcal{B}$ *mix, and Digest Forty days in* Igne Philosophorum: *So have you* Mercury, *not at all differing from the Natural,* But that is not fit for our work, as the Mineral is. If you have any understanding, this Caution may sufficiently instruct you.

CHAPTER XXXIX

OF THE PURIFICATION OF THE METALS AND MERCURY FOR OUR WORK.

I. This is a great and certain truth, that the Clean ought to be separated from the Unclean, for nothing can give that which it has not: For the pure substance is of one simple Essence, void of all Heterogeneity: But that which is impure and unclean, consists of Heterogene parts, is not simple, but compounded (to wit of pure and impure) and apt to putrifie and corrupt.

II. Therefore let nothing enter into your Composition, which is Alien or Foreign to the matter, (as all Impurity is) for nothing goes to the Composition of our Stone, that proceedeth not from it, neither in part nor in whole.

III. If any strange or foreign thing be mixed with it, it is immediately corrupted, and by that Corruption your Work becomes frustrate.

IV. The Citrine Bodies (as Sol, etc.) you must purge by Calcination or Cementation; and it is then purged or purified if it be fine and florid.

V. The metal being well cleansed, beat it into thin Plates or Leaves (as is Leaf Gold,) and reserve them for use.

VI. The White Liquor (as Mercury) contains two Superfluities, which must necessarily be removed from it, viz. Its foetid *Earthiness,* which hinders its Fusion: and its *Humidity,* which causes its flying.

VII. The Earthiness is thus removed. *Put it into a Marble or Wooden mortar, with its equal weight of pure fine and dry salt, and a little Vinegar: Grind all with the Pestle, till nothing of the matter appears, but the whole Salt becomes very black. Wash this whole matter with pure Water, till the Salt is dissolved, this filthy water decant, and put to the Mercury again as much more Salt and Vinegar, grinding it as before, and washing it with fair water, which work so often repeat till the water comes clear from it, and then the Mercury remains pure bright and clear like a Venice Looking Glass, and of a Coelestial Colour. Then strain it through a Linen Cloth three or four times doubled, two or three times (into a clean Glass Vessel) till it be dry.*

VIII. The proporation of the parts is as 24 to 1.

There are 24 Hours in a Natural Day, to which add
one, and it is 25. (To Wit, the Rising of the Sun).
To understand this, is Wisdom, as Geber saith.
Indeavour through the whole Work, to over-power the
Mercury in Commixtion.

IX. Rhasis saith, Those Bodies come nearest to
perfection, which contain most Argent Vive: He
farther saith, That the Philosophers hid nothing but
Weight and Measure, to wit, the Proportions of the
Ingredients, which is clear, for that none of them
all agree one with another therein: which causeth
great error.

X. Though the matters be well prepared and well
mixed, without the Proportions or Quantities of the
things be just, and according to the reason of the
Work, you will miss of the truth, or the end, and
lost all your Labour; you will not indeed bring
anything to perfection.

XI. And this is evident in the Examination: When
there is a Transmutation of the Body, or that the
Body is changed, then let it be put into the
Cineritium or Test, and then it will be consumed, or
otherwise remain; according as the proportions are
more or less than just; or just as they ought to be.

XII. If they be right and just, according to the
Reason of that, your Body will be incorruptible and
remain firm, without any loss, through all Essays
and Tryals: you can do nothing in this work without
the true knowledge of this thing, whose Foundation
is Natural matter, purity of substance, and right
Reason or proportion.

CHAPTER XL

OF THE CONJUNCTION OF THE PRINCIPLES, IN ORDER TO THIS
GREAT WORK.

I.   Euclid the Philosopher, and a man of great understanding, advises to work in nothing but in Sol and Mercury; which joyned together make the wonderful and admirable Philosophers Stone, as Rhasis saith: White and Red, both proceed from one Root; no other Bodies coming between them.

II.   But yet the Gold, wanting *Mercury,* is hindered from working according to his power. Therefore know that no Stone, nor Pearl, or other Forreign thing, but this our Stone, belongs to this work.

III. You must therefore Labour about the Dissolution of the Citrine Body, to reduce it into its first matter: for as *Rhasis* saith, *We dissolve Gold, that it may be reduced into its first* Nature *or matter, that is, into* MERCURY.

IV.   For being broken and made One, they have in themselves the whole Tincture both of the *Agent* and *Patient.* Wherefore saith Rhasis, make a Marriage (that is a Conjunction) between the RED Man and his

WHITE Wife, and you shall have the whole Secret.

V.   The same saith *Merlin:* If you Marry the White Woman to the Red Man, they will be Conjoyned and Imbrace one another, and become impregnated. By themselves they are Dissolved, and by themselves they bring forth what they have conceived, whereby, the two are made but one Body.

VI.   And truly our Dissolution, is only the reducing the hard Body into a liquid form, and into the Nature of Argent Vive, that the Saltness of the Sulphur may be diminished.

VII. Without our Brass then be Broken, Ground, and Gently and Prudently managed, till it will be reduced from its hard and Dense Body, into a thin and subtil Spirit, you labour in Vain.

VIII. And therefore in the *Speculum Alchymiae* it is said, the first work is the reducing the Body into Water, that is, into Mercury. And this the Philosophers called Dissolution, which is the Foundation of the whole Art.

IX.   This Dissolution makes the Body of an Evident Liquidity, and absolute Subtilty: and this is done by a gentle Grinding, and a soft and continued

Assation or Digestion.

X. Wherefore saith Rhasis, the work of making our Stone is, that the matter be put into its proper Vessel, and continually Decocted and Digested, until such time as it wholly Ascends, or Sublimes to the top thereof.

XI. This is declared in *Speculum Philosophorum*. The Philosophers Stone is converted from a vile thing, into a pretious Substance: for the SEMEN SOLARE is cast into the Matrix of Mercury, by Copulation or Conjunction, whereby in process of time they be made one.

XII. Also, that when it is Compounded with the like, and Mercurizated, then it shall be the Springing Bud. For the *Soul,* the *Spirit* and the *Tincture* may then be drawn out of them by the help of a gentle Fire.

XIII. Therefore saith Aristotle, the true matters or principles are not possible to be transformed or changed (by the most Learned in Alchymie) except they be reduced into their first matter.

XIV. And Geber saith, all ought to be made of Mercury only: for when *Sol* is reduced to its first

Original or Matter, by Mercury, then Nature embraceth Nature.

XV. And then it will be easie to draw out the Subtil and Spiritual Substance thereof: of which *Alkindus* saith, take the things from their Mines, and Exact or Subtilize them, and reduce them to their Roots, or first matter, which is *Lumen Luminum*.

XVI. And therefore, except you cast Out the *Redness* with the *Whiteness,* you will never come to the exalted glory of the *Redness*. For Rhasis saith, He that knows how to convert *Sol* into *Luna,* knows how to convert *Luna* into *Sol.*

XVII. Therefore saith *Pandophilus* in *Turba Philosophorum* he that prudently draws the Virtue or Power from Sol, and his Shadow, shall obtain a great Secret.

XVIII. Again it is said, without Sol, and his Shadow, no Tinging Virtue or Power is generated.

XIX. And whosoever it is that shall endeavour to make a Tinging or Colouring Tincture, without these things, and by any other means, he Errs, and goes astray from Truth, to his own hurt, loss and detriment.

CHAPTER XLI

OF THE VESSEL LUTE, CLOSING, AND TIMES OF THE PHILOSOPHICK WORK.

I. The Vessel for our Stone is but one, in which the whole Magistery or Elixir is performed and perfected; this is a Cucurbit, whose Bottom is round like an Egg, or an Urinal, smooth within, that it may Ascend and Descend the more easily, covered with a Limbeck round and smooth everywhere, and not very high, and whose Bottom is round also like an Egg.

II. Its largeness aught to be such; that the Medicine or matter may not fill above a fourth part of it; made of strong double Glass, clear and transparent, that you may see through it, all the Colours appertaining to, and appearing in the work; in which the Spirit moving continually, cannot pass or flie away.

III. Let it also be so closed, that as nothing can go out of it, so nothing can enter into it; as *Lucas* saith, *Lute the Vessell strongly with* LUTUMSAPIENTIAE, *that nothing may get in or go out of it.*

IV. For if the Flowers, or matter subliming, should breath out, or any strange Air or matter enter in, your work will be spoiled and lost.

V. And though the Philosophers oftentimes say, that the matter is to be put into the Vessel, and closed up fast, yet it is sufficient for the Operator, once to put the said matter in, once to close it up, and so to keep it even to the very perfection and finishing of the work. If these things be often repeated, the work will be spoiled.

VI. Therefore saith Rhasis, keep your Vessel continually close, encompassed with Dew, (which demonstrates what kind of Heat you are to use,) and so well Luted that none of the Flowers, or that which sublimes, may get out, or vanish in Vapor or Fume.

VII. And in Speculum Alchymiae it is said, Let the Philosophers Stone remain shut within the Vessel strongly, until such time, that it has drunk up the Humidity; and let it be nourished with a continual Heat till it becomes White.

VIII. Also another Philosopher in his Breveloquium saith, as there are three things in a natural Egg, viz, the Shell, the White, and the Yolk, so likewise there are three things corresponding to the Philosophers Stone, the Glass Vessel, the White Liquor, and the Citrine Body.

IX. And as of the Yolk and White, with a little Heat, a Bird is made, (the Shell being whole, until the coming forth or Hatching of the Chicken:) so is it in the work of the Philosophers Stone. Of the Citrine Body, and White Liquor, with a temperate or gentle Heat is made the Avie Hermetis, or Philosophers Bird.

X. The Vessel being well and perfectly closed, and never so much as once opened till the perfection or end of the work: so that you see the Vessel is to be kept close, that the Spirit may not get out and vanish.

XI. Therefore saith Rhasis, Keep thy Vessel and its junctures close and firm, for the Conservation of the Spirit. And another saith, close thy Vessel well, and as you are not to cease from the work, (or let it cool,) so neither are you to make too much haste, (neither by too great a heat, nor too soon opening of it.)

XII. You must take special care that the Humidity (which is the Spirit) gets not out of the Vessel;

for then you will have nothing but a Dead Body remaining, and the work will come to nothing.

XIII. *Socrates* saith, Grind it with most sharp Vinegar, till it grows thick and be careful that the Vinegar be not turned into fume, and perish.

CHAPTER XLII

OF THE PHILOSOPHERS FIRE, THE KINDS AND GOVERNMENT
THEREOF.

I. The Philosophers have described in their Books a two-fold Fire, a moist and a dry.

II. The *moist Fire* they called the warm *Horse Belly;* in the which, so long as the Humidity remains, the Heat is retained; but the Humidity being Consumed, the Heat vanishes and ceases, which Heat being small seldom lasts above five or six days: but it may be Conserved and renewed, by casting upon it many times Urine mixt with Salt.

III. Of this Fire speaks *Philares* the Philosopher: The property of the fire of the Horse Belly, is, not to destroy with it dryness the Oyl, but augments it with its humidity, whereas other fire would be apt to consume it.

IV. *Senior* the Philosopher saith, Dig a Sepulchre and bury the WOMAN with her MAN, or Husband in Horse-dung (or Balneo of the same heat) until such time as they be intimately conjoyned or united.

V. *Altudonus* the Philosopher saith likewise, you

must bide your Medicine in Horse Dung, which is the fire of the Philosophers, for this Dung is hot, moist, and dark, having a humidity in itself, and an excellent light (or Whiteness).

VI. There is no other fire comparable to it in the World, excepting only the natural heat of a Man, or Womans Body.)

VII.   This is a Secret. The Vapour of the Sea not burned, the Blood of Man, and the Blood of the Grape is our Red Fire.

VIII. The *Dry Fire,* is the Fire of the Bodies themselves and the Inflammability of everything able to be burned: Now the government of these Fires is thus:

IX. The Medicine of the White ought to be put into the moist fire, until the Complement of the Whiteness shall appear in the Vessel. For a gentle fire is the conservation of the Humidity.

X. Therefore saith *Pandolphus,* You are to understand that the Body is to be dissolved with the Spirit; with which they are mixed by an easie and gentle decoction, so that the Body may be spiritualized by it.

XI. *Ascanius* also saith: A gentle fire gives health, but too much or great a heat will not conserve or unite the Elements, but on the contrary divide them, waste the humidity, and destroy the whole work.

XII. Therefore saith Rhasis, Be very diligent and careful in the sublimation and liquefaction of the matter, that you increase not your fire too much, whereby the water may ascend to the highest part of the Vessel: For then wanting a place of Refrigeration, it will stick fast there, whereby the Sulphur of the Elements will not be perfected.

XIII. For indeed in this work, it is necessary that they be many times elevated, or sublimed, and depressed again.

XIV. And the gentle or temperate Fire is that only which compleats the mixture, makes thick, and perfects the work.

XV. Therefore saith *Botulphus,* That gentle fire, which is the White fire of the Philosophers, is the greatest and most principal matter of the Operation of the Elements.

XVI. Rhasis also saith, Burn our Brass with a Gentle

Fire, such as is that of a Hen for the hatching of Eggs, until the Body be broken, and the Tincture extracted.

XVII. For with an easie decoction, the water is congealed, and the humidity which corrupteth, drawn out; and in drying, the burning is avoided.

XVIII. The happy prosecution of the whole work, consists in the exact temperament of the fire: Therefore beware of too much heat, lest you come to *solution* before the time, (viz., before the matter is ripe:) For that will bring you to despair of attaining the end of your hopes.

XIX. Wherefore saith he, Beware of too much fire, for if it be kindled before the time, the matter will be Red, before it comes to ripeness and perfection, (whereby it becomes like an Abort, or the unripe Fruit of the Womb: whereas it ought to be first White, then Red, like as the Fruits of a Tree, a Cherry is first White, then Red, when it comes to its perfection.)

XX. And that he might indigitate a certain time, (as it were) of Decoction, he saith, That the dissolution of the Body, and Coagulation or Congelation of the Spirit, ought to be done, by an

easie decoction in a gentle fire, and a moist
Putrefaction, for the space of one hundred and forty
Days.

XXI. To which *Orsolen* assents saying, In the
beginning of the mixture, you ought to mix the
Elements (being sincere and made pure, clean, and
rectified with a gentle fire) by a slow and natural
digestion, and to beware of too much fire, till you
know they are conjoyned.

XXII. *Bonellus* also saith, That by a Temperate and
Gentle heat continued, you must make the pure and
perfect Body.

CHAPTER XLII

OF THE AENIGMA'S OF PHILOSOPHERS,
THEIR DECEPTIONS,
AND PRECAUTIONS CONCERNING THE SAME.

I. You ought to put on Courage, Resolution and Constancy, in attempting this great work, lest you Err, and be deceived, sometimes following or doing one thing, and then another.

II. For the knowledge of this Art consisteth not in the multiplicity, or great number of things, but in Unity; Our Stone is but One, the matter is One, and the Vessel is One: The Government is One, and the disposition is One. The whole Art and Work thereof is One, and begins in One manner, and in One manner it is finished.

III. Notwithstanding the Philosophers have subtily delivered themselves, and clouded their instructions with Anigmatical and Typical Phrases and Words, to the end that their Art might not only be hidden and so continued, but also be had in the greater Veneration.

IV. Thus they advise to Decoct, to Commix, and to Conjoyn; to Sublime, to Bake, to Grind, and to

Congeal; to make Equal, to Putrefie, to make White, and to make Red; of all which things, the order, management and way of working is all one, which is only to Decoct.

V. And therefore to Grind is to Decoct, of which you are not to be weary, saith Rhasis: Digest continually, but not in haste (that is, not with too great a Fire;) cease not, or make no intermission in your work, follow not the Artifice of Sophisters, but pursue your Operation, to the Complement and perfection thereof.

VI. Also in the *Rosary* it is advised, to be cautious and watchful, lest your work prove dead or imperfect, and to continue it with a long Decoction. Close up well thy Vessel, and pursue to the end.

VII. For there is no Generation of things, but by Putrefaction, by keeping Out the Air, and a continual internal motion, with an equal and gentle Heat.

VIII. Remember when you are in your work, all the Signs and Appearances which arise in every Decoction, for they are necessary to be known and understood in order to the perfecting the matter.

IX. You must be sure to be incessant and continual in your Operation, with a gentle Fire, to the appearing of the perfect Whiteness, which cannot be if you open the Vessel, and let out the Spirit.

X. From whence it is Evident, that if you manage your matter ill, or your Fire be too great, it ought to be extinguished: Therefore saith Rhasis, pursue your business incessantly, beware of instability of mind, and too great expectations, by a too hasty and precipitate pursuit, lest you lost your End.

XI. But as another Philosopher saith, Digest, and Digest again, and be not weary: The most exquisite and industrious Artist, can never attain to perfection by too much haste, but only by a long and continual Decoction and Digestion, (for so Nature works, and Art must in some measure imitate Nature.)

CHAPTER XLIV

OF THE VARIOUS SIGNS APPEARING IN EVERY OPERATION.

I. This then is the thing, that the Vessel with the Medicine be put into a moist Fire; to wit, that the middle or one half of the Vessel be in a moist Fire (or Balneo, of equal Heat with Horse—Dung,) and the other half out of the Fire, that you may daily look into it.

II. And in about the space of Forty Days, the superficies or upper part of the Medicine will appear black as melted Pitch: and this is the Sign, that the Citrine Body is truly converted into Mercury.

III. Therefore saith Bonellus, when you see the blackness of the Water to appear, be assured that the (Citrine) Body is made Liquid: The same thing saith Rhasis; the Disposition or Operation of our Stone is One, which is, that it be put into its Vessel and carefully Decocted and Digested, till such time as the whole Body be Dissolved and Ascended.

IV. And in another place he saith, continue it upon a temperate or gentle Balneo, so long till it be perfected Dissolved into Water, and made impalpable, and that the whole Tincture be extracted into the blackness, which is the Sign of its dissolution.

V. Lucas also assureth us, that when we see the blackness of the water in all things to appear, that then the Body is dissolved, or made Liquid.

VI. This blackness the Philosophers called the first Conjunction; for then the Male and Female are joyned together, and it is the Sign of perfect mixtion.

VII. Yet notwithstanding, the whole Tincture is not drawn out together; but it goes out every day, by little and little, until by a great length of time, it is perfectly extracted, and made compleat.

VIII. And that part of the Body which is dissolved, ever Ascends or Rises to the Top, above all the other undissoived Matter which remains yet at Bottom.

IX. Therefore saith Avicen, That which is spiritual in the Vessel Ascends up to the Top of the Matter, and that which is yet gross and thick, remains in the Bottom of the Vessel.

X. This blackness is called among the Philosophers

by many Names, to wit, *The Fires, the Soul, a Cloud, the RavensHead, a Coal, Our Oyl, Aqua vitae, the Tincture of Redness, the Shadow of the Sun, Black Brass, Water of Sulphur;* and by many other Names.

XI. And this Blackness is that which conjoyneth the Body with the Spirit.

XII. Then saith Rhasis, The Government of the Fire being observed for the space of Forty Days, both (to wit the White Liquor, and the Citrine Body) are made a Permanent or fixt Water, covered over with blackness; which blackness (if rightly ordered) cometh to its perfection in Forty Days space.

XIII. Of which another Philosopher saith; so long as the obscure blackness appeareth the WOMAN hath the Dominion; and this is the first Conception or strength of our Stone: For if it be not first *Black,* it shall never be either *White* or *Red*.

XIV. Avicen saith, That Heat causeth blackness first, in a moist Body; then the humidity being consumed, it putteth off or leseth its blackness; and as the Heat encreaseth (or is continued) so it grows white.

XV. Lastly, by a more inward Heat, it is Calcin'd

into Ashes, as the Philosophers teach.

XVI. In the first Decoction (which is called Putrefaction) Our Stone is made all Black, to wit, a Black Earth, by the drawing out of its Humidity; and in that Blackness, the Whiteness is hidden.

XVII. And when the Humidity is reverted upon the Blackness again, and by a continued soft and gentle Digestion is made fixed with its Earth, then it becomes White.

XVIII. In this Whiteness, the Redness is hidden; and when it is Decocted and Digested by augmentation (and continuance) of the Fire, that Earth is changed into Redness, as we shall hereafter teach.

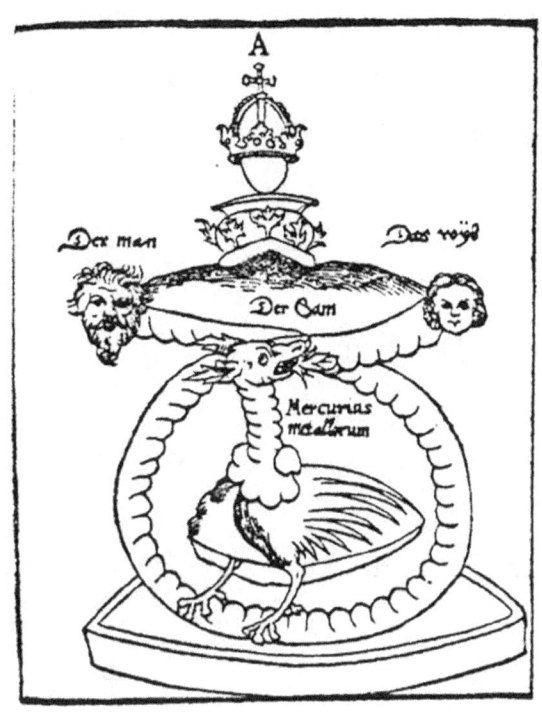

The Seed

CHAPTER XLV

OF THE EDUCATION OF THE WHITENESS OUT OF THE
BLACKNESS
OR BLACK MATTER,

I. Now let us revolve to the Black matter in its
Vessel, (not so much as once opened, but)
continually closed: Let this Vessel I say, stand
continually in the moist fire, till such time as the
White Colour appears, like to a white moist Salt.

II.   The Colour is called by the Philosophers *Arsenick,* and *Sal Armoniack;* and some others call it, *The thing without which no profit is to be had in the work.*

III.   But inward whiteness appearing in the Work, then is there a perfect Conjunction, and Copulation, of the Bodies in this Stone, which is indissoluble and then is fulfilled that saying of *Hermes,* The thing which is above, is as that which is beneath, and that which is beneath, is as that which is above, to perform the Mystery of this matter.

IV. *Phares* saith, Seeing the Whiteness appearing above in the Vessel, you may be certain, that in the Whiteness, the Redness lies hid; but before it becomes White, you will find many Colours to appear.

V. Therefore saith *Diomedes,* Decoct the Male and the (Female or) Vapour together, until such time as they shall become one dry Body; for except they be dry, the divers or various Colours will not appear.

VI. For it will ever be black, whilst that humidity or moisture has the dominion; but if that be once wasted, then it emits divers Colours, after many and several ways.

VII. And many times it shall be changed from Colour to Colour, till such time as it comes to the fixed Whiteness.

VIII. *Synon* saith, *All the Colours of the World will appear in it when the Black humidity be dryed up.*

IX. But value none of these Colours, for they be not the true Tincture; yea many times it becomes Citrine and Reddish; and many times it is dryed, and becomes liquid again, before the Whiteness will appear.

X. Now all this while the Spirit is not perfectly joyned with the Body, nor will it be joyned or fixed but in the White Colour: *Astanus* saith, Between the White and the Red appear all Colours, even to the utmost imagination.

XI. For the varieties of which the Philosophers have given various Names, and almost innumerable; some for obscuring it, and some for envy sake.

XII. The cause of the appearance of such variety of Colours in the Operation of your Medicine, is from the extention of the blackness; for as much as Blackness and Whiteness be the extreme Colours, all the other Colours are but means between them.

XIII. Therefore as often as any degree or portion of Blackness descends, so often another and another Colour appears, until it comes to Whiteness.

XIV. Now concerning the Ascending and Descending of the Medicine, Hermes saith, It ascends from the Earth into Heaven, and again descends from Heaven to the Earth, whereby it may receive both the superiour strength and the inferiour.

XV. Moreover this you are to observe, that if between the Blackness and the Whiteness, there should appear the Red or Citrine Colour, you are not to look upon it or esteem it, for it is not fixt but will vanish away.

XVI. There cannot indeed be any perfect and fixt Redness, without it be first White: Wherefore saith Rhasis, no Man can come from the first to the third, but by the second.

XVII. From whence it is evident, that Whiteness must always be first lookt for, (after the Blackness, and before the Redness,) for as much as it is the Complement of the whole Work.

XVIII. Then after this Whiteness appears, it shall not be changed into any true or stable Colour, but into the Red: Thus have we taught you to make the White; it remains now that we elucidate the Red.

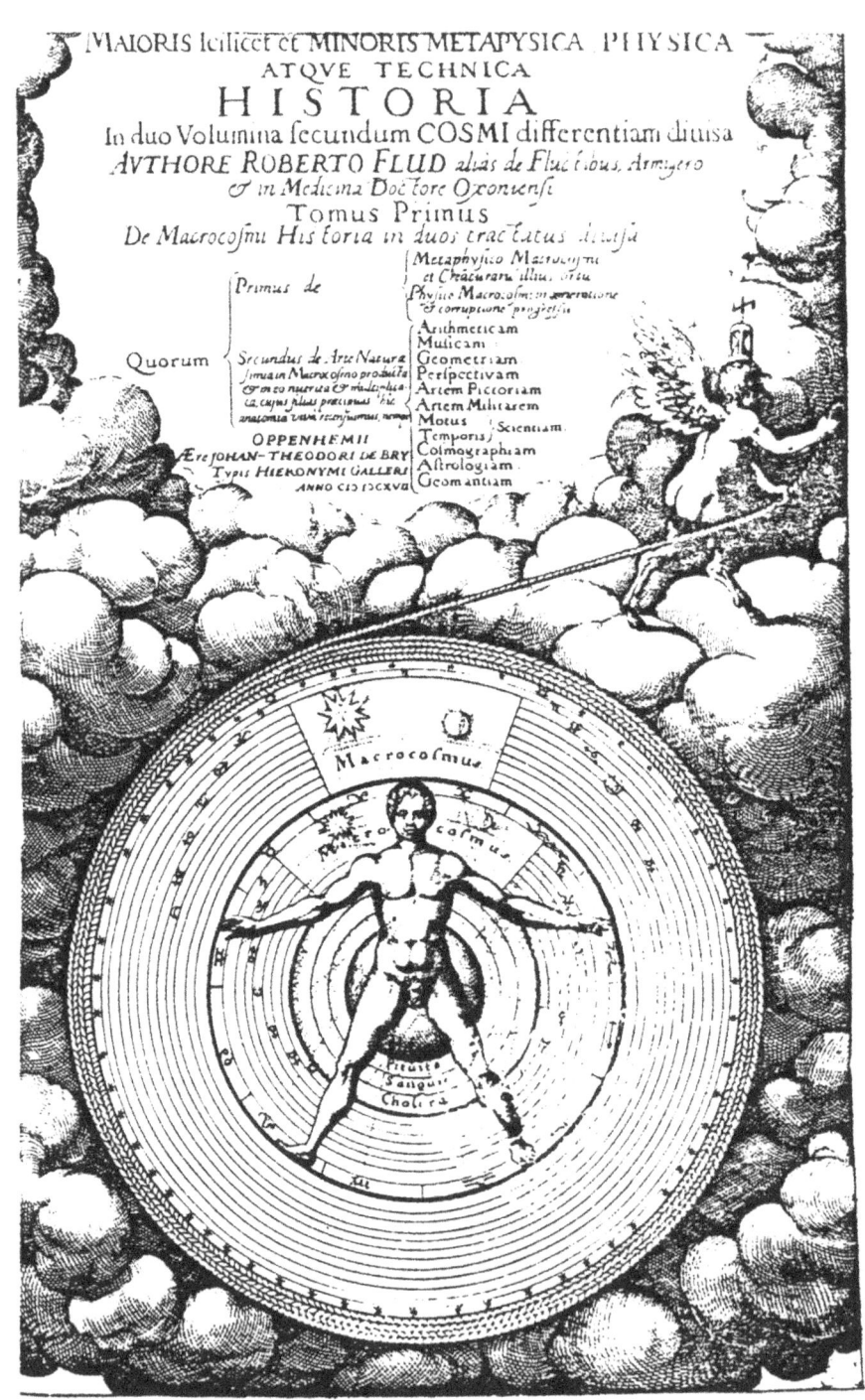

The Rosicrucian Correspondence between Macrocosm and Microcosm, Eternity and Time, illustrated by this engraved title page of Robert Fludd's principal work, de Bry, 1617.

CHAPTER XLVI

OF THE WAY AND MANNER HOW TO EDUCE THE RED TINCTURE OUT OF THE WHITE.

I. The matters then of the White and Red, among themselves, differ not in respect to the Essence: But the Red Elixir needs more subtilization, and longer digestion, and a hotter fire in the course of the Operation, than the White, because the end of the White work, is the beginning of the Red work; and that which is compleat in the one, is to be begun in the other.

II. Therefore without you make the White Elixir first, make the matter become first White, you can never come to the Red Elixir, that which is indeed the true Red: Which how it is to be performed we shall briefly shew.

III. The Medicine for the Red ought to be put into our moist fire, until the White Colour aforesaid appear, afterwards take out the Vessel from the fire, and put it into another pot with sifted Ashes made moist with water, to about half full, in which let it stand up to the middle thereof, making under the Earthen pot a temperate dry fire, and that continually.

IV. But the heat of this dry fire ought to be double at the least, to what it was before, or than the heat of the moist fire, by the help of this heat, the white Medicine receiveth the admirable Tincture of the Redness.

V. You cannot err if you continue the dry fire: Therefore *Rhasis* saith, *With a dry fire, and a dry Calcination decoct the dry matter, till such time as it becomes in Colour, like to Vermillion or Cinabar.*

VI. To the which you shall not afterwards put (to compleat it) either Water, or Oyl, or Vinegar, or any other thing.

VII. Decoct the Red Matter, or Medicine; the more red it is, the more worth it is; and the more decocted it is, the more red it is: Therefore that which is more decocted, is the more pretious and valuable.

VIII. Therefore you must burn it without fear in a dry fire, until such time as it is clothed with a most Glorious Red, or a pure Vermilion Colour.

IX. For which cause *Epistus* the Philosopher saith, *Decoct the White in a Red hot Furnace, until such time it be clothed with a purple Glory.* Do not cease, though the Redness be somewhat long, before it appears.

X. For as I have said, the fire being augmented, the first Colour of Whiteness, will change into Red: Also when the Citrine shall first appear, among those Colours, yet that Colour is not fixt.

XI. But not long after it, the Red Colour shall begin to appear, which ascending to the height, your Work will indeed be compleat.

XII. As Hermes saith in Turba, Between the Whiteness and the Redness, one Colour only appears, to wit, Citrine, but it changes from the less to the more.

XIII. *Maria* also saith, *When you have the true White, then follows the false and Citrine Colour;* and at last the Perfect Redness itself. This is the Glory and the beauty of the whole World.

CHAPTER XLVII

OF THE MULTIPLICATION, OR AUGMENTATION OF OUR MEDICINE, BY DISSOLUTION.

I. Our Medicine, or Elixir, is multiplyed after a twofold manner, viz. 1. By Dissolution. 2. By Fermentation.

II. By Dissolution, it is augmented two manner of ways, First, by a greater or more intense heat; Secondly, by Dew, or the heat of a *Balneum Roris*

III. The Dissolution of heat is, that you take the Medicine put into a glasen Vessel, or boil or decoct it in our moist fire for seven days or more, until the Medicine be dissolved into Water, which will be without much Trouble.

IV. The dissolution by Dew, or Balneum Roris, is, that you take the Glass Vessel with the Medicine in it, and hang it in a Brazen or Copper Pot, with a narrow Mouth, in which there must be water boyling, the Mouth of the Vessel being in the mean Season shut, that the Ascending Vapours of the boyling water may, dissolve the Medicine.

V. But Note, that the boyling water ought not to

touch the Glass Vessel, which contains the Medicine, by three or four Inches, and this Dissolution possibly may be done in two or three days.

VI. After the Medicine is dissolved, take it from the Fire, and let it cool, to be fixed, to be congealed, and to be made hard or dryed, and so let it be dissolved many times; for so much the often it is dissolved, so much the more strong, and the more perfect it shall be.

VII. Therefore Bonellus saith, When the AEs, *Brass*, or Laten is burned, and this burning many times reiterated, it is made better than it was: and this Solution is the Subtilization of the Medicine, and the Sublimation of the Virtues thereof.

VIII. So that the oftner it is sublimed and made subtil, so much the more Virtue it shall receive; and the more penetrative shall the Tincture be made, and more plentiful in quantity; and the more perfect it is, the more it shall transmute.

IX. In the Fourth Distillation then, it shall receive such a Virtue and Tincture that one part shall be able to transmute a thousand parts of the cleansed Metal into fine Gold or Silver, better than that which is Generated in the Mines.

X. Therefore saith *Rhasis,* the goodness or excellency of the Multiplication hereof depends only on the Reiteration of the dissolution and fixation of the perfect Medicine.

XI. For so much the oftner the work is Reiterated, so much the more fruitful it will be, and so much the more augmented.

XII So much the oftner you sublime it, so much the more you increase it for every time it is augmented in Virtue, and Power, and Tincture, one more to be cast upon a thousand, at a second time upon ten thousand, at the third time upon one hundred thousand, at the fourth time upon a Million: And thus you may increase its Power by the number of the Reiterations, till it is almost infinite.

XIII. Therefore saith *Meredes* the Philosopher, *knew for certain, that the oftener the Matter or Stone is dissolved and congealed, the more absolutely and perfectly, the Spirit and Soul are conjoyned and retained.*

XIV. And for this cause, every time the Tincture is Multiplied, after a most admirable and unconceiveable manner.

CHAPTER XLVIII

OF THE AUGMENTATION OR MULTIPLICATION OF OUR MEDICINE BY FERMENTATION.

I.   Our Medicine is Multiplied by Fermentation; and the Ferment for the *White* is pure *Luna,* the Ferment for the *Red,* is pure fine *Sol.*

II.   Now cast one part of the Medicine upon twenty parts of the Ferment, and all shall become Medicine, Elixir, or Tincture: Put it on the Fire in a Glass Vessel, and seal it so that no Air go in or out, dissolve and subtilize it, as oft as you please, even as you did for making of the first Medicine.

III. And one part of this second Medicine, shall have as much Virtue and Power, as Ten parts of the former.

IV.   Therefore saith *Rhasis,* Now have we accomplished our Work by that which is hot and moist, and it is become equally temperate: and whatsoever is added or put to it, shall become of the same temperament and Virtue with it.

V.   You must then Conjoyn it, that it may Generate its like; yet you must not joyn it with any other

57

that it might convert it to the same, but only with that very same kind, of whole substance it was in the beginning.

VI. For in *Speculo Terrae Spiritualis,* it is written, that the Elixir is figured in the Body, from whence it was taken in the beginning, when it was to be dissolved

VII. That is to say, to dispose, Marry or Conjoyn that Earth revived, and in its Soul purified by commixtion of its first Body, from whence it took beginning.

VIII. Also in *Libro Gemmae Salutaris,* it is said, that the White work needs a White Ferment; which when it is made White, is White Ferment also: and when it is made Red, is the Ferment of Redness.

IX. And so the White Earth is Ferment of Ferment: for when it is Conjoyned with Luna; or shall be made a Medicine, it is to cast upon Mercury, and every imperfect Metalline Body, to be converted into Luna.

X. And to the Red, ought Sol to be joyned; and it will become a Medicine or Tincture, to project upon Mercury, or upon Luna.

XI. Rhasis also saith, You must now mix it with Argent Vive, White and Red, after their kind; and be so chained that it flies not away.

XII. Wherefore we command Argent Vive to be mixed with Argent Vive, until one clear water made of two Argent Vive's compounded together.

XIII. But you must not make the mixture of them, till each of them apart or separately be dissolved into water: and in the Conjunction of them, put a little of the matter upon much of the Body, viz. First upon four; and it shall become in a short time a fine Pouder, whose Tincture shall be White or Red.

XIV. This Pouder is the true and perfect Elixir or Tincture, and the Elixir or Tincture, is truly a simple Pouder.

XV. *Egidius* also saith, to Solution put Solution, and in dissolution put desiccation, viz, make it dry, putting all together to the fire.

XVI. Keep entire the fume or vapour, and take heed that nothing thereof flie out from it: Tarry by the Vessel and behold the wonders, how it changes from Colour to Colour, in less space than an hour's time, till such time as it comes to the Signs of Whiteness or Redness.

XVII. For it melts quickly in the Fire, and congeals in the Air. When the fume or vapor feels the force of the fire, the fire will penetrate into the Body, and the Spirit will become fixed, and the matter made dry, becoming a Body fixt and clear or pure, and either White or Red.

XVIII. This Pouder is the compleat and perfect Elixir or Tincture, now you may separate or take, if from the fire, and let it cool.

XIX. And first, part of it projected upon 1000 parts of any Metalline Body, transmutes it into fine Gold or Silver, according as your Elixir or Tincture is for the Red or the White.

XX. From what has been said, it is manifest and Evident, that if you do not congeal Argent Vive, making it to bear or endure the fire, and then conjoyning it with pure Silver, you shall never attain to the Whiteness.

XXI. And if you make not Argent Vive Red, and so as it may endure the greatest fire, and then conjoyn it with pure fine Gold, you shall never attain to the

Redness.

XXII. And by dissolution, viz, by Fermentation, your Medicine, Elixir, or Tincture, may be multiplied infinitely.

XXIII. Now you must understand that the Elixir or Tincture, gives fusion like Wax: for which cause saith Rhasis, Our Medicine ought of necessity to be of a subtle substance, and most pure, cleaving to Mercury, of its Nature, and of most easie and thin liquifaction, fusion, or melting, after the manner of water.

XXIV. Also in the Book, called *Omne datum Optimum,* it is said, when the Elixir is well prepared, it ought to be made liquid, that it may melt as Wax upon a Plate Red Fire-Hot, or upon Coals.

XXV. Now observe what you do in the White, the same you must do in the Red, for the work is all one: The same Operation that is in the one, is in the other, as well in multiplication as projection.

CHAPTER XLIX

OF THE DIFFERENCES OF THE MEDICINE AND PROPORTIONS USED IN PROJECTION.

I. Geber, the Arabian Prince, Alchymist, and Philosopher, in lib. 5. cap. 21. saith, That there are three orders of Medicines. The First Order, is of such Medicines, which being cast upon imperfect Bodies, takes not away their Corruption or Imperfection, but only give Tincture, which in Examination, flies away and vanishes.

II. The Second Order, is of such Medicines, which being cast upon Imperfect Bodies, tinge them (in examination) with a true Tincture, but take not away wholly their Corruption.

III. The Third Order, is of such Medicines, which being cast upon Imperfect Bodies, not only perfectly tinge them, but also take away all their Corruption and Impurities, making them incorrupt and perfect: of the first two kinds, or Orders of Medicines, we have nothing to say here; we now come to speak of the third.

IV. Let therefore this your perfect Medicine, or Elixir, be cast upon a thousand or more parts,

according to the number of times it has been
dissolved, sublimed, and made subtil: If you put on
too little, you must mend it by adding more;
otherwise the Virtue thereof will accomplish a
perfect Transmutation.

V. The Philosophers therefore made three
Proportions, divers manner of ways, but the best
proportion is this: Let one part be cast upon an
hundred parts of Mercury, cleansed from all its
Impurities; and it will all become Medicine, or
Elixir; and this is the second Medicine: which
projected upon a thousand parts, converts it all
into good Sol, or Luna.

VI. Cast one part of this second Medicine upon a hundred of Mercury prepared, and it will all become
Medicine, and this is the Third Medicine, or Elixir
of the third degree, which will project upon ten
thousand parts of another Body, and transmute it
wholly into fine Sol or Luna.

VII. Again, every part of this Third Medicine
being cast upon an hundred parts of prepared
Mercury, it will all become Medicine of the fourth
degree, and it will transmute ten hundred Thousand
times its own quantity of another Metal into fine
Sol or Luna, according as your fermentation was

made.

VIII. Now these second, third, and fourth Medicines may be so often dissolved, sublimed, and subtilizated, till they receive far greater virtues and powers, and may after the same manner be multiplyed infinitely.

IX. According to Rhasis, the proportion is thus to be computed. First, multiply Ten by Ten, and its product is an Hundred: Again 100 by 10, and the product is 1000; and a 100 by 10, and the product will be 10000.

X. And this 10000 being multiplyed by 10, produces an 100000; and thus by consequence you may augment it, till it comes to a number almost infinite.

XI. That is to say, put 1 upon 10, and that 10 upon an 100, and that 100 upon a 1000, and it shall multiply to, or produce an 100000; and so forward, in the same proportion.

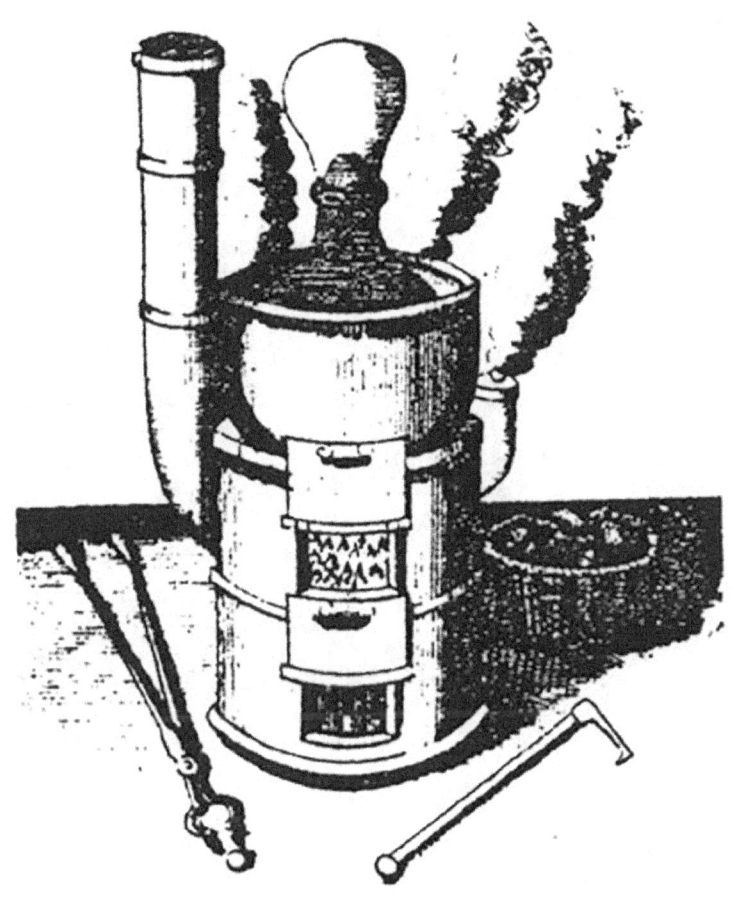

(ii) Vase of Hermes and Athanor.

A representation of the sealed vessel of Hermes and
the furnace used in the Great Work. From
the same.

CHAPTER L

OF PROJECTION, AND HOW IT IS TO BE PERFORMED UPON THE METALS.

I. Now the projection is after this manner to be done: Put the Body, or Metal upon the fire in a Crucible, and cast thereon the Elixir as aforesaid, moving, or stirring it well; and when it is melted, become liquid, and mixed with the Body, or with the Spirit, remove it from the fire, and you shall have fine Gold or Silver, according to what your Elixir was prepared from.

II. But here is to be noted, That by how much the more the Metalline Body is the easier to be melted, by so much the more shall the Medicine have power to enter into, and transmute it.

III. Therefore by so much as Mercury is more liquid than any other Body, by so much the more, the Medicine has power in being cast upon it, to wit, Mercury, to transmute it into fine Sol or Luna.

IV. And a greater quantity of it shall your Medicine transmute, give tincture to, and make perfect, than of any other Mineral Body.

V. The like is to be understood, to be performed in the same manner upon other Mineral Bodies, according as they are easie or hard to be fused or melted.

CHAPTER LI

OF THE COMPLEATMENT, OR PERFECTION OF THE WHOLE WORK.

I. And because prolixity is not pleasant, but induceth errour, and clouds the understanding, we shall now use much brevity, and shew the Complement of the whole work, the premises being well conceived.

II. It appears, that our Work is hidden in the Body of the Magnesia's, that is, in the Body of Sulphur; which is Sulphur of Sulphur; and in the Body of Mercury, which is Mercury of Mercury.

III. Therefore our Stone is from one thing only, as is aforesaid, and it is performed by one Act or Work, with decoction: and by one Disposition, or Operation, which is the changing of it first to Black, then to White, thirdly, to Red: and by one Projection, by which the whole Act and Work is finished.

IV. From henceforth, let all Pseudo-Chymists, and their Followers, cease from their vain Distillations, Sublimations, Conjunctions, CalcinatiOns, Dissolutions, Contritions, and such other like Vanities.

V. Let them cease from their deceiving, prating,
and pretending to any other Gold, than our Gold; or
any Sulphur, or any other Argent Vive than ours; or
any other Ablution or washing than what we have
taught.

VI. Which washing is made by means of the black
Colour, and is the cause of the White, and not a
washing made with hands.

VII. Let them not say, That there is any other
Dissolution than ours, or other Congelation than
that which is performed with an easie fire: or any
other Egg than that which we have spoken of by
similitude, and so called an Egg.

VIII. Or that there is any production of the
Philosophick matter from Vegetables, or from
Mankind, or from Brute Beasts, or Hare's Blood, and
such like, which may serve to this Work, lest
thereby you be deceived, and err, and fall short of
the end.

IX. But hear now what Rhasis saith, Look not upon
the multitude, or diversity of Names, which are dark
and obscure, they are chiefly given to the diversity

69

of Colours appearing in the Work.

X. Therefore whatever the Names be, and how many soever, yet conceive the matter or thing to be but one, and the work to be but one only.

XI. Lucas saith, Consider not the multitude of the Simples composing it, which the Philosophers have dictated, for the verity is but one only, in the which is the Spirit and Life sought after.

XII. And with this it is that we tinge and colour every Body, bringing them from their beginnings and smallness, to their compleat growth, and full perfection.

XIII. *Permenides* the Philosopher saith, It is a Stone, and yet no Stone, it is sulphur, and no Sulphur, It is Gold, and yet no Gold: It is also a Hen's Egg, a Toad, Man's Blood, Man's Hair, etc. by which Names it is called only to hide the Mystery.

XIV. Then he adds, *O thou most happy, let not these words, nor other the like trouble thee, for by them understand our Sulphur and our Mercury.*

XV. If you that are searchers into this Science, understand these words and things which we have written, you are happy, yea, thrice happy: If you understood not what we have said, God himself has hidden the thing from you.

XVI. Therefore blame not the Philosophers but your selves; for if a just and faithful mind possessed your souls, God would doubtless reveal the verity to you.

XVII. And know, it is impossible for you to attain to this knowledge, unless you become sanctified in mind, and purified in soul, so as to be united to God, and to become one Spirit with him.

XVIII. When you shall appear thus before the Lord, he shall open to you the Gates of his Treasure, the like of which is not to be found in all the Earth.

XIX. Behold, I shew unto you the fear of the Lord, and the love of him with unfeigned obedience: Nothing shall be wanting to them that fear God, who are cloathed with the Excellency of his Holiness, to whom be rendred all Praise, Honour, and Glory to the Ages of Ages, Amen.

CHAPTER LII

THE PREFACE OR ENTRANCE INTO THIS WORK, AND THE DEFINITION OF THE ART.

I. After many ways and in divers manners, the Ancient Philosophers have through all their writings delivered themselves; and in Aenigmaes or Riddles, they have wholly Clouded and left shadowed to us, the most Noble Science, and as it were under a Veil of Desperation, have wholly denied Us the knowledge thereof, and that not without cause.

II. For which Reason sake, I here signifie (that you may the more firmly Establish your mind) I have in the following Chapters declared (more plainly than is taught in any other writings) the whole Art of the Transformation of Metals.

III. And if you often revolve these instructions in your minds, you will obtain the knowledge of the beginning, the middle, and the end of the Work; and you will see such a subtility of Wisdom, and, such a purity of matter which amply repleat your Soul, and fill you with Satisfactions.

IV. Now in the ancient Codes, many definitions of this Art are to be found, the meaning of which it behoves us to consider, *Hermes* Saith concerning this Art, *it is the secret Science of compounded Bodies, joyning together, (through the knowledge of the matter and its effects or Operations) the more pretious things one to another, and by a Natural Commission, to convert or transmute the same into a better kind.*

V. Another Defines it *thus, Alchymie is a Science teaching how to transmute all kinds of metals, one into another, and this by a proper Medicine,* as appears in many Books of the Philosophers.

VI. Wherefore, *Alchymie is the Art or Science, teaching how to make or generate a certain kind of Medicine, which is called the ELIXIR and which being projected upon Metals or imperfect Bodies, by thoroughly Tinging and fixing them, perfects them in the highest degree, even in the very moment of Projection.*

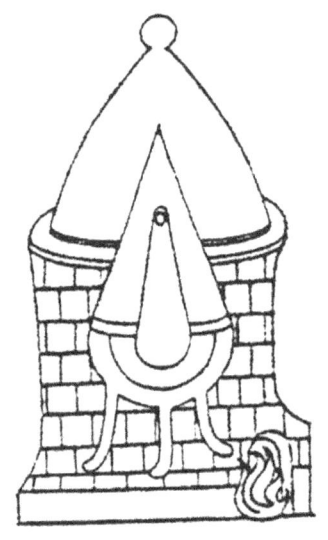

Athanor, from the Book of the Holy Trinity.

Alembic with retort.

CHAPTER LIV

OF THE NEAREST MATTER, OUT OF WHICH THE ELIXIR IS DRAWN OR MADE.

I.   In what we have already declared, we have spoken sufficiently of the generation of Metals; now we apply ourselves to the choice and perfecting of those which are imperfect.

II. Out of what has been said, it appears, that from *Argent Vive,* and Sulphur, all the Metals are generated; and how with their impurities and uncleanness they are corrupted: And therefore whatsoever matter does adhere to any Metal, which is not of its *Prima Materia,* or does not belong to its composition, it is to be rejected and cast away.

III. For that no Foreign matter, which is not composed of the aforesaid two principles, viz. *Argent Vive* and *Sulphur,* can either perfect a Metal, or make any new transmutation thereof.

IV. This is also to be wondered at, that even some wise prudent Persons; should lay the Foundation or whole matter of the Elixir, in the Animal or Vegetable Kingdoms, which are so infinitely remote from the thing, whilst they may find Mineral much nearer akin to the Work and Design.

V. It is not indeed to be at all believed, that any of the Philosophers, should place the Art, or Secret in such remote things, where there is not the least congruity or similitude of Natures.

VI. But out of the two aforesaid things, (viz. Argent Vive, and Sulphur) all Metals are made: and nothing does adhere to them, nor is conjoyned with them, nor can transmute them, except that which arises from the same Root or Principles.

VII. And therefore we say, that it behoves you indeed and in truth, to take Argent Vive, and Sulphur, for the matter of our Stone, not Argent Vive, by itself alone; nor Sulphur by itself alone; for neither of them alone can generate Metals: but from a commixtion of both, divers Metals are variously generated, as also many Minerals.

VIII. Therefore from a commixtion of them both, our matter of the Stone remains to be chosen, which is most excellent and deeply hidden: from which Mineral matter, that which is yet nearer and more akin thereto is to be made. And this very thing itself, we attain to the knowledge of, by a diligent and accurate search and enquiry.

IX. Take then this our Matter, chosen as you may think out of Vegetables, and from thence, first make Argent Vive and Sulphur, by a long decoction: But since Nature has given us Argent Vive and Sulphur, we are excused from those things, and their too tedious Operation: nor yet out of them could you produce the said Prima Materia.

X. And if you should seek for it in Animals, as in humane Blood, Hair, Urine, Dung, Hens Eggs, or any other things proceeding from Animals; from these you should also make Argent Vive and Sulphur, by such a like long decoction; but in all these things, you would miss of the Prima Materia, as you did before in Vegetables.

XI. If also you should make choice of the middle Minerals, such as are all the kinds of Magnesia's Marchasities, Tutias, Vitriols, Alums, Borax, Salt, and many others of like Rank, you must from these make Argent Vive and Sulphur, by long Boyling, without which, you would proceed in Vain; yet in Operating upon these things also, you would Err.

XII. If also you should make choice of some one of the seven Spirits by itself, as alone of Argent Vive, or of Sulphur alone, or of Argent Vive and one of the two Sulphurs, or of *Sulphur Vive,* or Auripigment, i.e. *Arsenicum Citrinum,* or of the Red alone, or its compeer, you would do nothing.

XIII. Because Nature does nothing, except there be a just or proportional mixtion of the two principles; nor can we (for the same Reason) from the aforesaid Argent Vive and Sulphur, as they are in their own Nature, does anything.

XIV. Lastly, if we should chuse even the things themselves, be they what they will, we ought to mix them according to the just and true proportion, which humane ingeny is ignorant of, and then to decoct or boyl them together, till they are coagulated into a solid mass.

XV. And therefore we forbid the taking of those two, viz. Argent Vive, and Sulphur, as they are, or lie in their own proper Natures, and being also ignorant of the just proportion of Parts for the mixtion.

XVI. So that we must find out those Bodies in which we may find the aforesaid things or principles justly proportionated, coagulated, and joyned together in one, as their Natures require: which Secret is very warily to be kept.

CHAPTER LV

OF THE NEAREST MATTER OF OUR STONE, YET MORE PLAINLY,

I. Gold is a Body perfect, and Masculine, without any superfluity or diminution, and if the imperfect Bodies commixed with it by a sole Liquefaction, be perfected by it, it is in Order for the Elixir for the Red.

II. Silver also is a Body almost perfect, and Feminine, which if it be commixed with imperfect bodys, solely by a vulgar fusion, it shall make them nearly perfect, it is in Order for the Elixir for the White, which yet it is not, nor can be, because the Elixirs only are perfect.

III. Because if that perfection was perfectly commiscible with imperfect Bodies, yet would not the imperfect Body be compleated with the perfect Bodies, but rather their perfection would be diminished and destroyed by means of the said Imperfect Bodies.

IV. But if those which shall be more than perfect, in a Double, Quadruple, Centuple, or larger proportion of perfection, be mixed with the imperfect Bodies, they will indeed perfect them.

V. And because Nature always Operates after a simple manner, the perfection in these things is simple, and inseparable, and incommiscible; nor by this Art, are the imperfect things themselves (for the shortening the work) to be joyned with the Stone for the Ferment, nor may they then be reduced into their pristine State, when their exceeding Volatility exceeds the highest fixity.

VI. And because Gold is a body perfect, made of *Argent Vive,* Red, and clear, and of such a like *Sulphur,* we do not therefore chose it, for the near matter of the Stone for the Red elixir for that by reason it is simply so perfect, without any artificial purification, and so strongly Digested, and Decocted by a Natural Heat, we cannot so easily Operate upon it (nor upon Silver) with our Artificial Fire.

VII. And altho Nature may do something in Order to perfection, yet it does not know how throughly to cleanse, and is ignorant how to Purifie and perfect, because it works after a simple manner upon what it hath.

VIII. Wherefore, if we chuse Gold or Silver for the

matter of our Stone, we shall scarcely, or with
difficulty find out a Fire which will work upon
them.

And though we know the Fire, yet we may not be able
to attain to the intimate and inward opening of
their bodies, because of their firm compactedness,
or density of body and Natural composition:
therefore we refuse to take the first for the Red,
or the latter for the White.

X. When we shall find some thing or body extracted
from a pure matter, or a more pure *Sulphur* and
Argent Vive, above that which Nature has a little or
in some small Measure wrought or brought forth; then
possibly, by the help of our Fire, and manifold
experiences in this our Art, which an Ingenious and
continued Operation upon the matter, through a
congruous Decoction, Purification, Coloration, and
Fixation, we may attain and perfect the thing sought
after.

XI. Therefore that matter is to be chosen, in which
is a pure *Argent Vive,* clear, White, and also Red,
not yet brought to its compleatment or perfection,
but commixed equally and proportionally as it
requires, with such a like pure, clean, White and
Red *Sulphur.*

XII.   Which Matter is to be Coagulated into a solid Mass; and with Ingenuity and Prudence, by the help of our Artificial Fire, we may be able to accomplish, its intimate and perfect mundification, and attain the Purity of things, and to perform such a work or make such a body, as shall (after the compleatment of the Operation) be a Million of times stronger, and more pure and perfect than the simple bodies themselves, Decocted and made by a Natural heat.

XIII. Be therefore wise: for in this my subtle Discourse I have demonstrated plainly the matter of our Stone sought after, by manifest probation, to the truly Ingenious. Here you may taste of that which is most delectable, above all whatsoever the Philosophers have told you.

CHAPTER LVI

OF THE MANNER OF WORKING, AND OF THE MODERATION AND CONTINUING OF THE FIRE.

I. Now it is profitable that you may find out this Mystery, (if you will bend yourself to study and labour) and wholly casting off your Folly and Ignorance, become wise through the words which I speak; to the attainment of that true matter of the Philosophers, the Blessed Stone of the Wise, upon which the Operations of Alchymia are excercised; by which we endeavor to perfect the imperfect *Bodies* and thereby to make them better than the perfect.

II. And forasmuch as Nature has handed down to us Imperfect Bodies only with the perfect, it is our business to take the known matter, which we have declared in these Chapters, and by much Pains and Industry, through the help of our Art, to make it even more than perfect.

III. If you be ignorant of the manner of doing or working, What is the Cause? Truly because they you see not after what manner Nature (which sometimes perfects the Metals) frequently, or commonly operates.

IV. See you not, that in the mines, by the continual heat which is in those Mineral Mountains, the gross waters, are decocted, and so much inspissated, as therefore (through length of time) to be made *Argent Vive?*

V. And from the fat of the Earth, by the same decoction and heat, is generated Sulphur: and that by the same heat preserved and continued upon the same, from the aforesaid things, *(Viz., Argent Vive and Sulphur)* according to their Purities and Impurities, all the Metals are generated?

VI. And that Nature by a sole or only decoction, does make or bring to perfection the *perfect Bodies,* as well as all the Imperfect Bodies or Metals?

VII. O great madness! These things which I thus quarry about, would you bring to pass and perfect, by fantastick, strange and imperfect methods?

VIII. Now a certain Wise man saith: *You must necessarily err, who endeavour to outdo Nature; and to perfect the Metals, yea, more than perfect them, by new and foreign methods of Operation, invented in your dull and insensible Noddles.*

IX. And *that God has bestowed upon Nature a right*

method, a direct way, which is by continual decoction, which the insipid and Fools, through their ignorance, despise and scorn to imitate.

X. Also, *Fire and Azoth are sufficient for the.* Again, *Heat perfects all things, or all the Metals.* Moreover, *decoct, decoct, decoct, and be not weary. Make your fire gentle and soft, which may always burn and endure for many days with a constant equal; but let it not go out or decay, for then you will suffer the loss of all.*

XI. In another place, *continue thy work with patience.* And again, *Beat, or grind the matter seven times.* Then again, *You must know that with one matter, to wit,* the *Stone; by one way, to with, by decocting; and in one Vessel the whole Magistery is performed and perfected.*

XII. And in another place, *The matter is ground, with fire.* Also, *This work is much like, or may be compared to the Creation of Mankind.*

XIII. For, like as an Infant at first is nourished with food easily digested, or Milk; But for the strengthening of the Bones with stronger Food or Meat: So also this Magistery. At first you must feed it with a gentle Fire, by the force of which

Decoction, whatsoever you desire is to be done.

XIV. And although we always speak of a gentle fire; not withstanding you are always to understand it in this sense, that according to the Regimen, or method of the Operation, it is by degrees, or little by little augmented and increased, even unto the highest degree.

CHAPTER LVIII

OF THE QUALITY OF THE VESSELS AND FURNACES.

I. The limits, method, way and manner of working, we have even now determined; it follows that we treat next of the Vessel and Furance; after what manner, and with what matter it ought to be made.

II. When Nature, with a natural heat in the metallick Mines does decoct, it is true, it performs this decoction without any Vessel: Now though we propound to follow Nature in decocting; yet we cannot do it without a fit Vessel for that end.

III. Therefore let us see first, what kind of place that is, where Metals are generated. It is evidently manifest that they are produced in Mineral places, and that the generating heat is in the bottoms of the Mountains, where it is perdurable, and always equal, and whose nature is always to ascend; which in ascending continually desiccates everywhere where it passes, and coagulates the spissed or gross water hidden in the Bowels or Veins of the Earth of Mountain into *Argent Vive*.

IV. And if a mineral fat is in the same place, from such a like Earth, it shall be warmed, and gathered together in the Veins of the Earth, and it run through the Mountains, it becomes Sulphur.

V. And as you may see in the said veins of the said place, that Sulphur generated (as is said) of the fat of the Earth doth meet also with the *Argent Vive* (as aforesaid) in the said veins of the Earth, so also it produces a thickening, or inspissating of that Mineral Water.

VI. Also there, by the said heat in the bottoms of the Mountains, equal, and perdurable, through a very long space of time, there is generated divers and several Metals, according to the nature of the place or its diversity.

VII. This also you must know, that in the places where Minerals are found, there is always found a durable heat.

VIII. From these things, then, we ought always to note, that a Mineral Mountain is everywhere close to itself, externally; and also stoney: because if the heat should possibly get out no Metals would be generated.

IX. So also, if we intend to imitate Nature, we must necessarily have such a Furnace, as may have some semblance or likeness of a Mountain, not as to its magnitude, but as to its continued heat; so that the imposed fire, when it ascends may not find a way forth; but that the heat may reverberate back on the vessel, containing in itself (firmly closed up) the matter of the Stone.

X. Which vessel ought to be round, with a little neck, made of Glass, or some certain Earth, like in nature or closeness of body to Glass: Whose Mouth ought to be stopped or closed up with Bitumen, or other like Emplastick substance, or otherwise Hermetically sealed up, so as the least Vapour may not come forth.

XI. And like as in the Mines, the heat does not immediately touch the matter of the Sulphur and *Argent Vive,* because the Earth of the Mountain is everywhere between:

XII. So in like manner, the fire ought not immediately to touch the Vessel containing in itself, the matters aforesaid of our Stone: But in another closed Vessel in like manner *that* is to be put; so that the temperate heat may better, and more conveniently touch both above and below. and everywhere, the matter of our stone.
(A double Boiler using earth instead of water and

totally surrounding first vessel? hwn)

XIII. Upon which account *Aristotle* saith, *That Mercury, in the Light of Lights, is to be decocted in a threefold Vessal:* and that the Vessel be made of the most firm and pure Glass, or which is better, of Earth having in itself the nature of Glass. *(Of which kind is our late* China *and* Porcelain *Ware, brought to us out of* Persia, China, *and other places of the* East—Indies). *(Pyrex or Corning-ware seems best. hwn)*

CHAPTER LVIII

OF THE COLOURS, ACCIDENTAL AND ESSENTIAL, APPEARING IN THE WORK.

I. We have now taught you what the exquisite matter of the Stone is, and also the true manner of working; by what method and with what order the decoction of the Stone is to be performed, whence oftentimes arises divers and various colours in the Philosophick matter.

II. Concerning which Colours, a certan Wise Man Saith: *Quot colores, tot nomina; so many colors as it has, so many Names:* According to the diversity of Colours appearing in the operation, the Philosophers have given it several Names.

III. For which Reason, in the first operation of this our Stone, it is called *Putrefaction;* and our Stone is made black: For which reason saith a Philosopher, *When thou findest that black; know that in that blackness, whiteness is hidden;* and now it behoves us to extract that whiteness from its most subtil blackness.

IV. Now after the Putrefaction *(or blackness)* it grows red, but not with the true redness: of which one of the Philosophers saith, *It often grows red, and it often grows Citrine or Yellow; and it oftentimes melts, or grows liquid, and it is oftentimes coagulated, before the true Whiteness appears to perfection.*

V. Also it dissolves itself, coagulates itself, putrefies itself, tinges itself, or colours itself, mortifies itself, vivifies itself, denigrates or blackens itself, dealbates or whitens itself, and adorns itself in the red with the white.

VI. It is also made green: for which reason another saith: *Decoct it till you see the birth of the Greenness, or till the greenness is brought forth, which is the Soul thereof.* And another: Know *that the Soul does rule in the Greenness.*

VII. Also the colour of the Peacock appears before the *Whiteness;* for which cause, saith one: *Know that all the Colours which are in the World, or are possible to be thought of, appear before the* Whiteness *and then the* true Whiteness *follows.*

VIII. Of which a certain Philosopher saith: *But when the pure Stone is decocted, so long till the eye of the Fish (as it were) grows very bright; a profit or good may be expected from it; and then our Stone will be congealed into its roundness.*

IX. Another also saith: *When you shall find the Whiteness, Supereminent in the Vessel; be certain that in that Whiteness, the Redness is hidden; and then it behoves thee to extract it.*

X. Notwithstanding, decoct until the whole Redness be brought forth and perfected.

XI. For it is between the true Whiteness and the True Redness that a certain *Ash Colour* appears, of which we have spoken: after the *Whiteness* appears you cannot err, for by augmenting the fire you come to the *Ash Colour.*

XII. Of which another saith: *Slight or undervalue not the Ashes; for God will return them to thee liquid: and then at last the King shall be crowned with his red diadem,* NUTU DEI, *by the good pleasure of God.*

Creator Blesses the Globe

CHAPTER LIX

OF THE MANNER OF PROJECTION UPON ANY OF THE
IMPERFECT METALS.

I. I have perfectly compleated the end of the
promised Work, viz, of our great Magistery, for the
making of the most excellent Elixir, as well Red as
White: It now remains, that we shew the method, or
way of Projection, which is the compleatment of the

work, the long expected and much desired cause of rejoycing.

II. Now the True Red Elixir, tinges a pure and deep Citrine or Yellow, to infinity of parts, and it transmutes all Metals into most fine Gold.

III. The true White Elixir also, whitens to Infinity likewise; and it makes or tinges every Metal into a perfect Whiteness: But you must know, that one kind of Metal is much more remote, or far distant from perfection than some others are; and that some are much nearer to the said perfection than others.

IV. And although every Metal may be brought to perfection by the Elixir; yet those which are nearer to perfection, are easier, sooner and better reduced to that perfection, or transmuted into perfect Bodies than those that are more remote.

V. And when we have found a Metal, which is as it were, a kin or nearer to perfection, we are excused in some measure of making use of, or projecting upon those Metals, which are more remote therefrom.

VI. Now what Metals are remote from, and near to, perfection, and what are yet more near, and as it were a kin to the perfect Bodies, we have taught in these Chapters; in which if you be indeed wise you may plainly see, and truly determine what they are.

VII. And without doubt, he who is lawfully initiated into the Mysteries of this our Art; may be able through his own Ingenuity and Industry by studying this in my _Speculum Alchymiae_ to find out and know the true matter of the Stone: And he will know and understand well upon what Body, the Medicine or Magistery ought to be projected for perfection.

VIII. For the Masters of this Art, who have invented or found out the _Prima Materia,_ and the whole Mystery they have, I say, plainly demonstrated, and as it were, indigitated the direct way of working, and made all things naked and plain to us, when they say:

IX. _Nature contains Nature: Nature exceeds Nature, and Nature overcoming Nature does rejoyce, and is transmuted or changed into another Nature. And_ in another place, _every like doth rejoyce in its like; for that the likeness between things is said to be the cause of Sympathy or Friendship:_ of which many Philosophers have written notable things.

X. _Know then that the sol doth soon enter into its own Body_ but _with a Foreign or Strange Body, it can_ *never be joyned or United.* In another place: *If you shall endeavor to joyn it with a Forreign or Hetrogene Body, you shall wholly labour in vain.* Also, *The nearness of the Body to perfection makes a Transmutation the more Glorious.*

XI. For the Coporeal, by the power of the Operation of Nature, is made Incorporeal; and contrariwise the Incorporeal is made Corporeal; and in the compleatment, the spiritual Body is made wholly fixed.

XII. And because it is evidently manifest that the *Elixir* is Spiritual, and so very much exalted beyond its own Nature, as well for *the White,* as for *the Red:* It is no wonder, that it is not to be mixed with Bodies.

XIII. The Method, or way of Projection then is, that the Body of the Metal to be transmuted, be liquified or melted; and then that the Medicine or Elixir, be projected upon the melted Metal.

XIV. Moreover, you must Note, that this our Elixir, is of a mighty strong Power, and of great Force, for one part being projected upon a Million parts, or Ten Thousand Parts, and more, of the prepared Body, it does incontinently penetrate it, transfuse itself through the whole and transmute it.

XV. Wherefore I deliver to you a great and hidden Secret. Mix one part of this, our Elixir, with a thousand parts of a body near perfection; put all into a proper Vessel, inclosing it firmly; and then put it into a Furnace of Fixation first with a gentle fire, and then always augmenting the fire gradually for three days; so will they be inseparably conjoyned. This is a work of three days.

XVI. Then again, and lastly, take one part of this mixture, and project it upon a Thousand parts of another Body or Metal, (the nearer to perfection the better), so the whole will be a most fine and perfect Body, according to the kind, and according to your intended Work, whether for the White or for the Red.

XVII. And all this is but the work of one day, or rather, but of an hour, or of a moment: for which wonderful work, Praises be perpetually rendered to the Lord our God for Ever and Ever.

CHAPTER LX

A SHORT RECAPITULATION OF THE WHOLE WORK

I. Our Tincture then, is only generated out of the Mercury of the wise, which is called the *Prima Materia, Aqua Permanens, Acetum Philosophorum, Lac Virginis, Mercurius Corporalis,* with which nothing extraneous, Foreign or Alien is commixed, such as are Salts, Alums and Vitriols.

II. Because from this Mercury alone, the Virtue and Power of this our Magistery is generated: and it so resolves every (Metalline) Body, that it may be augmented or multiplied.

III. This our aforesaid Mercury is both the Root and the Tree, from whence many and almost Infinite Branches Spring and increase.

IV. The first work for the making of this Elixir, is sublimation, which is nothing else, but the subtilization of the matter of our Stone, by which it is cleansed from all its superfluities.

V. The fixed and volatile parts are not separated one from another, but they remain United, and are fixed together, till they both may have an easie fusion together in the fire.

VI. Take therefore our aforesaid *Mercury,* which in a sealed glass put into its hot bed, for one whole Philosophick Month which is 40 DAYS, till it begins in its own body to putrifie and be Coagulated, and all its humidity be confirmed in its self, and also converted into a black Earth.

VII. In this Sublimation is compleated the true separation of the four Elements.

VIII. Because the cold and watery Elements is changed into a hot and dry Earth, which is the *Head of the Crow,* the Mother of the remaining Elements.

IX. Thus our work is nothing else but a transmutation of Nature and a Conversion of the Elements.

X. The Spiritual is made Corporal, the Liquid is made thick, and the water is made Fire.

XI. Moreover, the black Earth is imbibed with its own water, and is dryed so long until it is sufficiently cleansed and brought on to Whiteness.

XII. Which then is called the White Earth foliated,

Sulphur of Nitre, *Sulphur of* Magnesia: and then there is a new conjunction of *Sol* and *Luna* and a resurrection of the Dead Body.

FINIS.

The Mirrour of Alchimy

Roger Bacon

"The mirror of alchimy, composed by the thrice-famous and learned fryer, Roger Bachon, sometimes fellow of Martin Colledge: and afterwards of Brasen-nose Colledge in Oxenforde. Also a most excellent and learned discourse of the admirable force and efficacie of art and nature, written by the same author. With certaine other treatises of the like argument."

LONDON.
Printed by Thomas Creede
for Richard Olive.

1597.

The Preface.

In times past the Philosophers spake after divers and sundrie manners throughout their writings, with that as it were in a riddle and cloudie voice, they have left unto us a certaine most excellent and noble science, but altogether obscure, and without all hope utterly denied, and that not without good cause. Wherefore I would advise thee, that above all other bookes, thou shouldest firmly fixe thy mind upon these seven Chapters, containing in them the transmutation of metals, and often call to mind the beginning, middle, and end of the same, wherein thou shalt find such subtilitie, that thy mind shall be fully contented therewith.

The Mirrour of Alchimy,

composed by the famous Fryer,
Roger Bachon,
sometime fellow of Martin College,
and Brasennose College in Oxford.

CHAP. I. Of the Definitions of Alchimy.

In many ancient Bookes there are found many definitions of this Art, the intentions whereof we must consider in this Chapter. For Hermes saith of this Science: Alchimy is a Corporal Science simply composed of one and by one, naturally conjoyning things more precious, by knowledge and effect, and converting them by a natural commixtion into a better kind. A certain other saith: Alchimy is a Science, teaching how to transforme any kind of metal into another: and that by a proper medicine, as it appeareth by many Philosophers Bookes. Alchimy therefore is a science teaching how to make and compound a certaine medicine, which is called Elixir, the which when it is cast upon metals or imperfect bodies, doth fully perfect them in the verie projection.

CHAP. II.

Of the naturall principles, and procreation of Minerals.

Secondly, I will perfectly declare the naturall principles & procreations of Minerals: where first it is to be noted, that the naturall principles in the mines, are Argent-vive, and Sulphur. All metals and minerals, whereof there be sundrie and divers kinds, are begotten of these two: but I must tell you, that nature alwaies intendeth and striveth to the perfection of Gold: but many accidents coming between, change the metals, as it is evidently to be seene in divers of the Philosophers bookes. For according to the puritie and impuritie of the two aforesaid principles, Argent-vive, and Sulphur, pure, and impure metals are ingendred: to wit, Gold, Silver, Steele, Leade, Copper, and Iron: of whose nature, that is to say, puritie, and impuritie, or unclean superfluitie and defect, give eare to that which followeth.

Of the nature of Gold.

Gold is a perfect body, engendered of Argentvive pure, fixed, cleare, red, and of Sulphur cleane, fixed, red, not burning, and it wanteth nothing.

Of the nature of Silver.

Silver is a body, cleane, pure, and almost perfect, begotten of Argent-vive, pure, almost fixed, cleare, and white, & of such alike *Sulphur:* It wanteth nothing, save a little fixation, colour, and weight.

Of the nature of Steele.

Steele is a body cleane, imperfect, engendred of Argent-vive pure, fixed & not fixed cleare, white outwardly, but red inwardly, and of the like Sulphur. It wanteth only decoction or digestion.

Of the nature of Leade.

Leade is an unclean and imperfect bodie, engendred of Argent-vive impure, not fixed, earthy, drossie, somewhat white outwardly, and red inwardly, and of such a Sulphur in part burning. It wanteth puritie, fixation, colour, and fiering.

Of the nature of Copper.

Copper is an unclean and imperfect bodie, engendred of Argent-vive, impure, not fixed, earthy, burning, red not cleare, and of the like Sulphur. It wanteth purity, fixation, and weight: and hath too much of an impure colour, and earthinesse not burning.

Of the nature of Iron.

Iron is an unclean and imperfect body, engendred of Argent-vive impure, too much fixed, earthy, burning, white and red not cleare, and of the like Sulphur: It wanteth fusion, puritie, and weight: It hath too much fixed unclean Sulphur, and burning earthinesse. That which hath been spoken, every Alchimist must diligently observe.

CHAP. III.

Out of what things the matter of Elixir must be more nearly extracted.

The generation of metals, as well perfect, as imperfect, is sufficiently declared by that which hath been already spoken. Now let us return to the imperfect matter that must be chosen and made perfect. Seeing that by the former Chapters we have been taught, that all metals are engendred of Argent-vive and Sulphur, and how that their impuritie and uncleannesse doth corrupt, and that nothing may be mingled with metals which hath not beene made or sprung from them, it remaineth cleane enough, that no strange thing which hath not his original from these two, is able to perfect them, or to make a change and new transmutation of them: so that it is to be wondred at, that any wise man should set his mind upon living creatures, or vegetables which are far off, when there be minerals to be found nigh enough: neither may we in any wise thinke, that any of the Philosophers placed the Art in the said remote things, except it were by way of comparison: but of the asoresaid two, all metals are made, neither doth anything cleave unto them, or is joined with them, nor yet changeth them, but that which is of them, and so of right we must take Argent-vive and Sulphur for the matter of our stone:

Neither doth Argent-vive by itself alone, nor Sulphur by itself alone, beget any metal, but of the commixtion of them both, divers metals and minerals are diversely brought forth. Our matter therefore must be chosen of the commixtion of them both: but our final secrete is most excellent, and most hidden, to wit, of what minerall thing that is more neere then others, it should be made: and in making choice hereof, we must be very warie. I put the case then, yet our matter were first of all drawne out of vegetables, (of which sort are herbs, trees, and whatsoever springeth out of the earth) here we must first make Argent-vive & Sulphur, by a long decoction, from which things, and their operation we are excused: for nature herself offereth unto us Argent-vive and Sulphur. And if we should draw it from living creatures (of which sort is man's blood, haire, urine, excrements, hens eggs, and what else proceede from living creatures) we must likewise out of them extract Argent-vive and Sulphur by decoction, from which we are freed, as we were before. Or if we should choose it out of middle minerals (of which sort are all kindes of Magnesia, Marchasites, of Tutia, Coppers, Allums, Baurach, Salts, and many others) we should likewise, as afore, extract Argent-vive and Sulphur by decoction, from which as from the former, we are also excused. And if we should take one of the seven spirits by itself, as Argent-vive, or Sulphur alone, or Argent vive and one of the two Sulphurs, or Sulphur-vive, or Auripigment, or Citrine Arsenicum, or red alone, or the like: we should never effect it, because sith nature doth never perfect anything without equal commixtion of both, neither can we: from these therefore, as from the foresaide Argent-vive and Sulphur in their nature we are excused. Finally, if we should choose them, we should mixe everything as it is, according to a due proportion, which no man knoweth, and afterward decoct it to coagulation, into a solid lump: and therefore we are excused from receiving both of them in their proper nature: to wit, Argent-vive and Sulphur, seeing we know not their proportion, and that we may meete with bodies, wherein we shall find the saide things proportioned, coagulated & gathered together, after a due manner. Keep this secret more secretly. Gold is a perfect masculine bodie, without any superfluitie or diminution: and if it should perfect imperfect bodies mingled with it by melting only, it should be Elixir to red. Silver is also a body almost perfect, and feminine, which if it should almost perfect imperfect bodyes by his common melting only, it should be Elixir to white, which it is not, nor cannot be, because they only are perfect. And if this perfection might be mixed with the imperfect, the imperfect should not be perfected with the perfect, but rather their perfections should be diminished by the imperfect, & become imperfect. But if they were more than perfect, either in a two-fold, four-fold, hundred-fold, or larger proportion, they might then well perfect the imperfect. And forasmuch as nature doth alwaies work simply, the perfection which is in them is simple, inseparable, & incommiscible, neither may they by art be put in the stone, for serment to shorten the worke, and so brought to their former state, because the most volatile doth overcome the most fixt. And for that gold is a perfect body, consisting of Argent-vive, red and cleare, & of such a Sulphur, therfore we choose it not for the matter of our stone to the red Elixir, because it is so simply perfect, without artificiall mundification, & so strongly digested and sod with a natural heate, that with our artificiall fire, we are scarcely able to worke on gold or Silver. And though nature dooth perfect anything, yet she cannot throughly mundifie, or perfect and purifie it, because she simply worketh on that which she hath. If therefore we should choose gold or Silver for the matter of the stone, we should hard and scantly find fire working in them. And although we are not ignoranr of the fire, yet could we not come to the through mundification & perfection of it, by reason of his most firme knitting together, and naturall composition: we are therefore excused for taking the first too red, or the second too white, seeing we may find out a thing or somebody of as cleane, or rather more cleane Sulphur & Argent-vive, on which nature hath wrought little or nothing at all, which with our artificiall fire, & experience of our art, we are able to bring unto his due concoction, mundification, colour and fixation, continuing our ingenious labour upon it. There must therefore be such a matter chosen, wherein there is Argent-vive, cleane, pure, cleare, white & red, not fully compleat, but equally and proportionably commixt after a due manner with ye like Sulphur, & congealed into a solide masse, that by our wisdome and discretion, and by our artificiall fire, we may attain unto the uttermost cleannesse of it, and the puritie of the same, and bring it to that passe, that after the worke ended, it might be a thousand thousand times more strong and perfect, then the simple bodies themselves, decoct by their naturall heate. Be therefore wise: for it thou shalt be subtile and wittie in my Chapters (wherein by manifest prose I have laid open the matter of the stone easie to be knowne) thou shalt taste of that delightfull thing, wherein the whole intention of the Philosophers is placed.

CHAP. IIII.

Of the manner of working, and of moderating, and continuing the fire.

I hope ere this time thou hast already found out by the words alreadie spoken (if thou beest not most dull, ignorant, and foolish) the certaine matter of the learned Philosophers blessed stone, whereon Alchimy worketh, whilest we indevour to perfect the imperfect, and that with things more than perfect. And for that nature hath delivered vs the imperfect only with the perfect it is our part to make the matter (in the former Chapters declared unto us) more than perfect by our artificiall labour. And if we know not the manner of working, what is the cause that we do not see how nature (which of long time hath perfected metals) doth continually work? Do we not see, that in the Mines through the continuall heate that is in the mountaines thereof, the grosnesse of water is so decocted & thickened, that in continuance of time it becommeth Argent-vive? And that of the fatnesse of the earth through the same heate and decoction, Sulphur is engendred? And that through the same heate without intermission continued in them, all metals are ingendred of them according to their puritie and impuritie? And that nature doth by decoction alone perfect or make al metals, as well perfect as imperfect? O extreame madnesse! What, I pray you, constraines you to seeke to perfect the foresaide things by strange melancholicall and fantasticall regiments? As one sayth: Wo to you that will overcome nature, and make metals more than perfect by a new regiment, or worke sprung from your owne senselesse braines. God hath given to nature a straite way, to wit, continuall concoction, and you like fooles despise it, or else know it not. Againe, fire and Azot, are sufficient for thee. And in another place, Heat perfecteth all things. And elsewhere, seeth, seeth, seeth, and be not wearie. And in another place, let thy fire be gentle, & easie, which being alwayes equall, may continue burning: and let it not increase, for if it do, thou shalt suffer great loss. And in another place, Know thou that in one thing, to wit, the stone, by one way, to wit, decoction, and in one vessel the whole mastery is performed. And in another place, patiently, and continually, and in another place, grinde it seven times. And in another place, It is ground with fire. And in another place, this worke is verie like to the creation of man: for as the Infant in the beginning is nourished with light meates, but the bones being strengthened with stronger: so this masterie also, first it must have an easie fire, whereby we must always worke in every essence of decoction. And though we alwayes speake of a gentle fire, yet in truth, we think that in governing the worke, the fire must alwayes by little and little be increased and augmented unto the end.

CHAP. V.

Of the qualitie of the Vessell and Furnace.

The meanes and manner of working, we have alreadie
determined: now we are to speake of the Vessell and
Furnace, in what sort, and of what things they must
be made. Whereas nature by a naturall fire decocteth
the metals in the Mines, she denieth the like
decoction to be made without a vessell fitte for it.
And if we purpose to immitate nature in concocting,
wherefore do we reiect her vessell? Let vs first of
all therefore, see in what place the generation of
metals is made. It doth evidently appeare in the
places of Minerals, that in the bottom of the
mountaine there is heate continually alike, the
nature whereof is always to ascend, and in the
ascention it alwayes drieth up, and coagulateth the
thicker or grosser water hidden in the belly, or
veines of the earth, or mountaine, into Argent-vive.
And if the minerall fatness of the same place
arising out of the earth, be gathered warme together
in the veines of the earth, it runneth through the
mountain, & becommeth Sulphur. And as a man may see
in the foresaid veines of that place, that Sulphur
engendred of the fatnesse of the earth (as is before
touched) meeteth with the Argent-vive (as it is also
written) in the veines of the earth, and begetteth
the thicknesse of the minerall water. There, through

the continual equall heate in the mountaine, in long
processe of time diverse metals are engendred,
according to the diversitie of the place. And in
these Minerall places, you shall find a continuall
heate. For this cause we are of right to note, that
the externall minerall mountaine is everywhere shut
up within it selfe, and stonie: for if the heate
might issue out, there should never be engendred any
mettall. If therefore we intend to immitate nature,
we must needes have such a furnace like unto the
Mountaines, not in greatnesse, but in continual
heate, so that the fire put in, when it ascendeth,
may find no vent: but that the heat may beat upon
the vessell being close shutte, containing in it the
matter of the stone: which vessell must be round,
with a small necke, made of glasse or some earth,
representing the nature or close knitting together
of glasse: the mouth whereof must be signed or
sealed with a covering of the same matter, or with
lute. And as in the mines, yet heat doth not
immediately touch the matter of Sulphur and Argent-
vive, because the earth of the mountain commeth
everywhere between: So this fire must not
immediately touch the vessell, containing the matter
of the foresaide things in it, but it must be put
into another vessell, shut close in the like manner,
that so the temperate heate may touch the matter
above and beneath, and where ere it be, more aptly

and fitly: whereupon Aristotle sayth, in the light of lights, that Mercurie is to be concocted in a threefold vessell, and that the vessell must be of most hard Glasse, or (which is better) of earth possessing the nature of Glasse.

CHAP. VI.

Of the accidental and essential colours appearing in the work.

The matter of the stone thus ended, thou shalt know the certaine manner of working, by what manner and regiment, the stone is often changed in decoction into diverse colours. Wherupon one saith, So many colours, so many names. According to the diverse colours appearing in the worke, the names likewise were varied by the Philosophers: whereon, in the first operation of our stone, it is called putrifaction, and our stone is made blacke: whereof one saith, When thou findest it blacke, know that in that blacknesse whitenesse is hidden, and thou must extract the same from his most subtile blacknes. But after putrefaction it waxeth red, not with a true rednesse, of which one saith: It is often red, and often of a citrine colour, it often melteth, and is often coagulated, before true whitenesse. And it dissolueth itselfe, it coagulateth itselfe, it putrifieth itselfe, it coloureth itself, it mortifieth itselfe, it quickneth itselfe, it maketh itselfe blacke, it maketh itselfe white, it maketh itselfe red. It is also greene: whereon another sayth, Concoct it, till it appeare greene unto thee, and that is the soule. And another, Know, that in that greene his soule beareth dominion. There appeares also before whitenesse the peacocks colour, whereon one saith thus. Know thou that all the colours in the world, or yet may be imagined, appeare before whitenesse, and afterward true whitenesse followeth. Whereof one sayth: When it hath bin decocted pure and clean, that it shineth like the eyes of fishes, then are we to expect his utilitie, and by that time the stone is congealed rounde. And another sayth: When thou shalt find whitenesse a top in the glasse, be assured that in that whitenesse, rednesse is hidden: and this thou must extract: but concoct it while it become all red: for betweene true whitenesse and true rednesse, there is a certaine ash-colour: of which it is said. After whitenesse, thou canst not erre, for increasing the fire, thou shalt come to an ash-colour: of which another saith: Do not set light by the ashes, for God shall give it thee molten: and then at the last the King is invested with a red crowne by the will of God.

CHAP. VII.

How to make projection of the medicine upon any imperfect body.

I have largely accomplisht my promise of that great masterie, for making the most excellent Elixir, red and white. For conclusion, we are to treate of the manner of projection, which is the accomplishment of the work, the desired & expected joy. The red Elixir doth turne into a citrine colour infinitely, and changeth all metals into pure gold. And the white Elixir doth instantly whiten, and bringeth every mettall to perfect whitenesse. But we know that one mettall is farther off from perfection then another, & one more neere then another. And although every mettall may by Elixir be reduced to perfection, neverthelesse the neerest are more easily, speedily, and perfectly reduced, then those which are far distant. And when we meete with a mettall that is neere to perfection, we are thereby excused from many that are farre off. And as for the metals which of them be neere, and which farre off, which of them I say be neerest to perfection, if thou be wise and discreete, thou shalt find to be plainely and truely set out in my Chapters. And without doubt, he that is so quick sighted in this my Mirrour, that by his own industry he can find out the true matter, he doth full well know upon what body the medicine is to be projected to bring it to perfection. For the forerunners of this Art, who have founde it out by their philosophie, do point out with their finger the direct & plain way, when they say: Nature, containeth nature: Nature overcommeth nature: & Nature meeting with her nature, exceedingly rejoyceth, and is changed into other natures. And in another place, Every like rejoiceth in his like: for likenesse is saide to be the cause of friendship, whereof many Philosophers have left a notable secret, Know thou that the soule doth quickly enter into his body, which may by no meanes be joined to another body. And in another place. The soule doth quickly enter into his own body, which if thou goest about to joyne with another body, thou shalt lose thy labour: for the neerenesse it selfe is more cleare. And because corporeal things in this regiment are made incorporeal, & contrariwise things incorporeal corporeal, and in the shutting up of the worke, the whole body is made a spirituall fixt thing: and because also that spirituall Elixir evidently, whether white or red, is so greatly prepared and decocted beyonde his nature, it is no marvel that it cannot be mixed with a body, on which it is projected, being only melted. It is also a hard matter to project it on a thousand thousand and more, and incontinently to penetrate and transmute them. I will therefore nowe deliver unto you a great and hidden secret. One part is to be mixed with a thousand of the next body, & let all this be surely put into a fit vessell, and sette it in a furnace of fixation, first with a lent fire, and afterwardes increasing the fire for three dayes, till they be inseparably joined together, and this is a worke of three dayes: then againe and finally, every part hereof by it selfe, must be projected upon another thousand parts of any neere body: and this is a worke of one day, or one houre, or a moment, for which our wonderfull God is eternally to be praised.

Here endeth the Mirror of Alchimy, composed by the most learned Philosopher, Roger Bacon.

Finis.

The Smaragdine Table

of

Hermes, Trismegistus of Alchimy.

The wordes of the secrets of Hermes, which were written in a Smaragdine Table, and found betweene his hands in an obscure vaute, wherein his body lay buried. It is true without leasing, certain and most true. That which is beneath is like that which is above: & that which is above, is like that which is beneath, to worke the miracles of one thing. And as all things have proceeded from one, by the meditation of one, so all things have sprung from this one thing by adaptation. His father is the sun, his mother is the moon, the wind bore it in her belly. The earth is his nurse. The father of all the telesme of this world is here. His force and power is perfect, if it be turned into earth. Thou shalt seperate the earth from the fire, the thin from the thicke, and that gently with great discretion. It ascendeth from the Earth into Heaven: and againe it descendeth into the earth, and receiveth the power of the superiours and inferiours: so shalt thou have the glorie of the whole worlde. All obscuritie therefore shall flie away from thee. This is the mightie power of all power, for it shall overcome every subtile thing, and pearce through every solide thing. So was the worlde created. Here shall be miraculous adaptations, whereof this is the meane. Therefore am I called Hermes Trismegistus, or the thrice great Interpreter: having three parts of the Philosophy of the whole world. That which I have spoken of the operation of the Sunne, is finished.

Here endeth the Table of Hermes.

A briefe Commentarie of Hortulanus the Philosopher, upon the Smaragdine Table of Hermes of Alchimy.

The prayer of Hortulanus.

Laude, honour, power and glorie, be given to thee, O Almightie Lorde God, with thy beloved sonne, our Lord Jesus Christ, and the holy Ghost, the comforter. O holy Trinitie, that art the only one God, perfect man, I give thee thanks that having the knowledge of the transitorie things of this worlde (least I should be provoked with the pleasures thereof) of thy abundant mercie thou hast taken me from it. But for so much as I have knowne many deceived in this art, that have not gone the right way, let it please thee, O Lord my God, that by the knowledge which thou hast given me, I may bring my deare friends from error, that when they shall perceive the truth, they may praise thy holy and glorious name, which is blessed forever.

Amen.

The Preface.

I Hortulanus, so called for the Gardens bordering upon the sea coast, wrapped in an Iacobin skinne, unworthy to be called a Disciple of Philosophie, moved with the love of my well beloved, do intend to make a true declaration of the words of Hermes, the Father of Philosophers, whose words, though that they be dark and obscure, yet have I truely expounded the whole operation and practise of the worke: for the obscuritie of the Philosophers in their speeches, doth nothing prevaile, where the doctrine of the holy spirit worketh.

CHAP. I.

That the Art of Alchimy is true and certaine.

The Philosopher saith. It is true, to wit, that the Arte of Alchimie is given unto us. Without leasing. This he saith in detestation of them that affirme this Art to be lying, that is, false. It is certaine, that is proved. For whatsoever is proved, is most certaine, And most true. For most true Gold is ingendered by Art: and he saith most true, in the superlative degree, because the Gold ingendered by this Art, excelleth all naturall gold in all proprieties, both medicinall and others.

CHAP. II.

That the Stone must be divided into two parts.

Consequentlie, he toucheth the operation of the stone, saying: That which is beneath, is as that which is above. And this he sayth, because the stone is divided into two principall parts by Art: Into the superiour part, that ascendeth up, and into the inferiour part, which remaineth beneath fixe and cleare: and yet these two parts agree in vertue: and therefore he sayeth, That which is above, is like that which is beneath. And this division is necessarie, To perpetrate the myracles of one thing, to wit, of the Stone: because the inferiour part is the Earth, which is called the Nurse, and Ferment: and the superiour part is the Soule, which quickeneth the whole Stone, and raiseth it up. Wherefore separation made, and conjunction celebrated, many miracles are effected in the secret worke of nature.

CHAP. III.

That the Stone hath in it the foure Elements.

And as all things have proceeded from one, by the
meditation of one. Here giveth he an example,
saying: as all things came from one, to wit, a
confused Globe, or masse, by meditation, that is the
cogitation and creation of one, that is the
omnipotent God: So all things have sprung, that is,
come out from this one thing that is, one confused
lumpe, by Adaptation, that is by the sole
commandment of God, and miracle. So our Stone is
borne, and come out of one confused masse,
containing in it the foure Elements, which is
created of God, and by his sole miracle our stone is
borne.

CHAP. IIII.

That the Stone hath Father and Mother, to wit, the Sun and Moon.

And as we see, that one living creature begetteth more living creatures like unto itselfe: so artificially Gold engendereth Gold, by vertue of multiplication of the foresaid stone. It followeth therefore, the Sunne is his father, that is, Philosophers Gold. And as in every naturall generation, there must be a fit and convenient receptacle, with a certaine consonancie of similitude to the father: so likewise in this artificiall generation, it is requisite that the Sunne have a fitte and consonant receptacle for his seede and tincture: and this is Philosophers Silver. And therefore it followes, the Moon is his mother.

CHAP. V.

That the conjunction of the parts of the stone is called Conception.

The which two, when they have mutuallic entertained each other in the conjunction of the Stone, the Stone conceiveth in the bellie of the winde: and this is it which afterwarde he sayeth: The winde carried it in his bellie. It is plaine, that the winde is the ayre, and the ayre is the life, and the life is the Soule. And I have already spoken of the soule, that it quickneth the whole stone. And so it behoveth, that the wind should carry and recarry the whole stone, and bring forth the masterie: and then it followeth, that it must receive nourishment of his nurse, that is the earth: and therefore the Philosopher saith, The earth is his Nurse: because that as the infant without receiving food from his nurse, should never come to years: so likewise our stone without the fermentation of his earth, should never be brought to effect: which said firmament, is called nourishment. For so it is begotten of one Father, with the conjunction of the Mother. Things, that is, sons like to the Father, if they want long decoction, shall be like to the Mother in whitenesse, and retaine the Fathers weight.

CHAP. VI.

That the Stone is perfect, if the Soule be fixt in the bodie.

It followeth afterward: The father of all the Telesme of the whole worlde is here: that is, in the worke of the stone is a finall way. And note, that the Philosopher calleth the worke, the Father of all the Telesme: that is, of all secret, or of all treasure Of the whole worlde: that is, of every stone found in the world, is here. As if he should say, Behold I shew it thee. Afterward the Philosopher saith, Wilt thou that I teach thee to know when the vertue of the Stone is perfect and compleate? To wit, when it is converted into his earth: and therefore he saith, His power is entire, that is, compleate and perfect, if it be turned into earth: that is, if the Soule of the stone (whereof we have made mention before: which Soule may be called the winde or ayre, wherein consisteth the whole life and vertue of the stone) be converted into the earth, to wit of the stone, and fixed: so that the whole substance of the Stone be so with his nurse, to wit earth, that the whole Stone be turned into ferment. As in making of bread a little leaven nourisheth and sermenteth a great deale of Paste: so will the Philosopher that our stone be so fermented, that it may be ferment to the multiplication of the stone.

CHAP. VII.

Of the mundification and cleansing of the stone.

Consequently, he teacheth how the Stone ought to be multiplied: but first he setteth downe the mundification of the stone, and the separation of the parts: saying, Thou shalt separate the earth from the fire, the thinne from the thicke, and that gently with great discretion. Gently, that is by little, and little, not violently, but wisely, to witte, in Philosophicall doung. Thou shalt separate, that is, dissolve: for dissolution is the separation of partes. The earth from the fire, the thinne from the thicke: that is, the lees and dregges, from the fire, the ayre, the water, and the whole substaunce of the Stone, so that the Stone may remaine most pure without all filth.

CHAP. VIII.

That the unfixed part of the Stone should exceed the fixed, and list it up.

The Stone thus prepared, is made fit for multiplication. And now he setteth downe his multiplication ct easie liquefaction, with a vertue to pierce as well into hard bodies, as soft, saying: It ascendeth from the earth into heaven, and again it descendeth into the earth. Here we must diligently note, that although our stone be divided in the first operation into foure partes, which are the foure Elements: notwithstanding, as we have alreadie saide, there are two principall parts of it. One which ascendeth upward, and is called unfixed, and another which remaineth below fixed, which is called earth, or firmament, which nourisheth and fermenteth the whole stone, as we have already said. But of the unfixed part we must have a great quantity, and give it to the stone (which is made most clean without all filth) so often by masterie that the whole stone be carried upward, sublimating & subtilating. And this is it which the Philosopher saith: It ascendeth from the earth into the heaven.

CHAP. IX.

How the volatile Stone may againe be fixed.

After all these things, this stone thus exalted, must be incerated with the Oyle that was extracted from it in the first operation, being called the water of the stone: and so often boyle it by sublimation, till by vertue of the firmentation of the earth exalted with it, the whole stone do againe descende from heaven into the earth, and remaine fixed and flowing. And this is it which the Philosopher sayth: It descendeth again into the earth, and so receiveth the vertue of the superiours by sublimation, and of the inferiours, by descention: that is, that which is corporall, is made spirituall by sublimation, and that which is spirituall, is made corporall by descension.

CHAP. X.

Of the fruit of the Art, and efficacie of the Stone.

So shalt thou have the glorie of the whole world.
That is, this stone thus compounded, thou shalt
possesse the glorie of this world. Therefore all
obscuritie shall flie from thee: that is, all want
and sicknesse, because the stone thus made, cureth
every disease. Here is the mightie power of all
power. For there is no comparison of other powers of
this world, to the power of the stone. For it shall
overcome every subtil thing, and shall pearce
through every solide thing. It shall overcome, that
is, by overcomming, it shall convert quicke Mercury,
that is subtile, congealing it: and it shall pearce
through other hard, solide, and compact bodies.

CHAP. XI.

That this worke imitateth the Creation of the worlde.

He giveth us also an example of the composition of his Stone, saying, So was the world created. That is, like as the world was created, so is our stone composed. For in the beginning, the whole world and all that is therein, was a confused Masse or Chaos (as is above saide) but afterward by the workmanship of the soveraigne Creator, this masse was divided into the source elements, wonderfully separated and rectified, through which separation, divers things were created: so likewise may divers things be made by ordering our worke, through the separation of the divers elements from divers bodies. Here shall be wonderfull adaptations that is, Is thou shalt separate the elements, there shall be admirable compositions, fitte for our worke in the composition of our Stone, by the elements rectified: thereof, to wit, of which wonderfull things fit for this: the meanes, to wit, to proceede by, is here.

CHAP. XII.

An enigmaticall insinuation what the matter of the Stone shoulde be.

Therefore am I called Hermes Trismegistus. Now that he hath declared the composition of the Stone, he teacheth us after a secret manner, whereof the Stone is made: first naming himself, to the end that his schollers (who should hereafter attaine to this science) might have his name in continuall remembrance: and then he toucheth the matter saying: Having three parts of the Philosophie of the whole world: because that whatsoever is in the worlde, having matter & forme, is compounded of the foure Elements: hence is it, that there are so infinite parts of the world, all which he divideth into three principall partes, Minerall, Vegetable, & Animall: of which jointly, or severally, he had the true knowledge in the worke of the Sunne: for which cause he faith, Having three parts of the Philosophic of the whole world, which parts are contained in one Stone, to wit, Philosophers Mercurie.

CHAP. XIII.

Why the Stone is said to be perfect.

For this cause is the Stone said to be perfect, because it hath in it the nature of Minerals, Vegetables, and Animals: for the stone is three, and one having foure natures, to wit, the soure elements, & three colours, black, white, and red. It is also called a graine of corne, which if it die not, remaineth without fruit: but if it do die (as is above said) when it is joined in conjunction, it bringeth forth much fruite, the afore named operations being accomplished. Thus curteous reader, if thou know the operation of the Stone, I have told thee the truth: but if thou art ignorant thereof, I have said nothing. That which I have spoken of the operation of the Sunne is finished: that is, that which hath beene spoken of the operation of the stone, of the three colours, and foure natures, existing and being in one only thing, namely in the Philosophers Mercurie, is fulfilled.

Thus endeth the Commentarie of Hortulanus, upon the Smaragdine table of Hermes, the father of Philosophers.

The Booke of the Secrets of Alchimie, composed by Galid the sonne of Iazich, translated out of Hebrew into Arabick, and out of Arabick into Latine, and out of Latin into English.

The Preface of the difficultie of the Art.

Thankes be giuen to God the Creator of all things, who hath conducted us, beautified us, instructed us, and given us knowledge and understanding: Except the Lorde should keep and guide vs, we should be like vagabonds, without guide or teacher: yea, we should know nothing in the world, unlesse he taught us: that is, the beginning, and knowledge itselfe of all things, by his power and goodness over his people. He directeth and instructeth whom he will, and with mercie reduceth into the way of justice: for he hath sent his messengers into the darke places, and made plaine the wayes, and with his mercy replenished such as love him. Know brother, that this our mastery and honourable office of the secret Stone, is a secret of the secrets of God, which he hath concealed from his people, neither would he reveale it to any, save to those, who like sonnes have faythfully deserved it, knowing both his goodnesse and greatnesse: for to him that desireth a secret of God, this secret masterie is more necessary than any

143

other. And those wise men who have attained to the knowledge herof, have concealed part thereof, and part thereof they have revealed: for so have I found my wise predecessors agreeing in this point in their worthie bookes, whereby thou shalt know that my disciple Musa, (more honorable in my eyes then all other) hath diligently studied their bookes, & labored much in the worke of the mastery, wherein he hath been greatly troubled, & much perplexed, not knowing the natures of things belonging to this work: the explanation whereof, and direction wherein, he hath humbly begged at my handes: yet I would afford him no answere therein, nor determine it, but commanded him to reade over the Philosophers bookes, & therin to seeke yet which he craved of me, & he going his way, read above a hundred bookes, as he found them even the true and secret bookes of noble Philosophers: but in them he could not find that which he desired: so he remained astonished, & almost distracted, though by the space of a yeare he continually sought it. If therefore my scholler Musa (that hath deserved to be accounted among ye Philosophers) have beene so doubtfull in the composition hereof, and that this hath happened unto him: what shall the ignorant and unlearned do, that understandeth not the nature of things, nor is acquainted with their complections? But when I behelde this in my choysest and dearest disciple, moved with pitty and compassion toward him, or rather by the will and appointment of God, I made this book at the houre of my death, wherin I have pretermitted many things, that my predecessors have made mention of in their bookes: and againe, I have touched some things which they concealed, & would by no meanes open & discover: yea, I have expounded and laide open certain things, that they have hidden under dark & figurative speeches. And this my book I have called the Secrets of Alchimy: in which I have spoken of whatsoever is necessarie, to him that is studious of this Art or masterie, in a language befitting his sense & understanding. And I have named foure masteries far greater and better, then other Philosophers have done: of which number is Elixir, one Mineral, the other Animall: but the other two are minerals, and not the one Elixir: whose office is to washe that, which they call the bodies: and another is to make gold of Azotvive, whose composition or generation, is according to the generationor order of generation in the mines, being in the heart and bowels of the earth. And these foure masteries or works, the Philosophers have declared in their bookes of the composition of this mastery: but they want much: neither would they shew the operation of it in their bookes: and though by chance he found it out, yet could he not understand it: so that he found out nothing that was more

145

troublesome to him. I will therefore in this my booke declare it, together with the manner how to make it: but let him that will reade it, first learne Geometry, and her measures, that so he may rightly frame his furnaces, not passing a meane, either by excesse or defect: and withall, he must know the quantitie of his fire, and the forme of the vessell fit for his worke. Moreover, lette him consider what is the ground-worke and beginning of the mastery, being to it, as the matrice is to living creatures, which are fashioned in the wombe, and therin receive their creation & nourishment: for if the thing of this mastery find not that which is convenient for it, the worke is marred, and the workmen shall not find that which they looke for, neither shal the thing itself be brought to the effect of generation: for where one cannot meete with the cause of generation, or the roote, and heate itselfe, it will fall out, that the labour shall be lost, and the worke nought worth. The like mischiefe will happen in respect of weight, which if it be not aright in the compound, the partes of the same nature, passing their boundes by augmentation, or diminution, the propertie of the compound is destroyed, & the effect thereof void and without fruit, whereof I will give you an example. Do not you see that in Sope (with which cloathes are washed cleane and made white) there is this property if it be rightly made, by reason of equalitie, & one proportion, which participate in length and breadth? whereupon through this participation they agree, and then it appeareth, because it was truely made, and so the vertue which before lay hid, is nowe made known, which they call a property, being the vertue of washing engendred in the compound: but when the gravity of the compound passeth his bounds, either by addition or diminution, ye vertue itself breaketh the limits of equality, & becometh contrary, according to ye distemperance of the compound. And this thou must understand to happen in the composition of our mastery.

CHAP I.

Of the foure Masteries, or principall works of the Art, to wit, solution, congelation, albification, and rubification.

Now begin I to speake of the great worke which they call Alchimy, wherein I will confirme my woordes, without concealing ought, or keeping backe anything, save that which is not convenient to be uttered or named. We say then that the great work containeth in it foure masteries (as the Philosophers before vs have affirmed) that is to say, to dissolue, to congeale, to make white and red. And these foure quantities are partakers, whereof two of them are partakers betweene themselues, and so likewise are the other two. And either of these double quantities hath another quantity partaker, which is a greater quantity partaker after these two. I understand by these quantities, the quantitie of the natures, and weight of the medicines which are orderly dissolved and congealed, wherin neither addition nor diminution have any place. But these two, to wit, solution and congelation, shall be in one operation, and shall make but one worke, and that before composition: but after composition, their works shall be divers. And this solution and congelation which we have spoken of, are the solution of the bodie, and the congelation of the Spirite, and they are two, yet have but one operation. For the Spirites are not congealed, except the bodies be dissolved, is likewise the bodies is not dissolved, unlesse the spirit be congealed: & when the soule & the body are joined together, either of them worketh in his companion made like unto him: as for example, when water is put to earth, it striveth to dissolve the earth by the moisture, vertue and propertie which it hath, making it more subtile then it was before, and bringing it to be like itselfe: for the water was more subtile then the earth: and thus doth the soule worke in the bodie, and after the same manner is the water thickened with the earth, and becommeth like unto the earth in thicknesse, for the earth is more thicke then the water. And thou must know that betweene the solution of the bodie, and congelation of the spirit, there is no distance of time or diverse work, as though one should be without the other, as there is no difference of time in the conjunction of the earth, and water, that one might be knowne & discerned from the other in their operations: but they have both one instant, and one fact, and one and the same worke containeth them both at once before composition: I say before composition, least he that shall read this booke, and heare the names of resolution and congelation, shoulde suppose it to be the composition which the Philosophers entreat of, for so he should fowly erre in his worke and judgement: because composition in this worke or masterie, is a conjunction or marriage of the congealed spirit, with the dissolved bodie, and this conjunction or passion is upon the fire. For heate is his nourishment, and the soule forsaketh not the bodie, neither is it otherwise knit unto it, then by the alteration of both from their owne vertue and properties, and after the conversion of their natures: and this is the solution and congelation, which the Philosophers first spake of: which neverthelesse they have hidden in their subtile discourses with darke & obscure words, that so they might alienate and estrange the mind of the reader from the true understanding thereof: where of thou maist take this for an example. Annoynt the leafe with poyson, and ye shall approve there by the beginning of the worke and mastery of the same. And againe, labour the strong bodies with one solution, til either of them be turned to his subtilitie. So likewise in these following, except ye convert the bodies into such subtilitie that they may be impalpable, ye shall not find that ye looke for: and if you have not ground them, returne backe to worke till they be ground, and made subtill: which if you do, you shall have your wish. And many other such sayings have they of the same matter. The which none that ever proved this Art could understand, til he hath had a plaine demonstration thereof, the former doubt being removed. And in like manner have they spoken of that composition, which is after solution & congelation. And afterward they have said, that Composition is not perfect without marriage, and putrifaction: yet againe they teach solution, congelation, division, marriage, putrifaction, and composition, because composition is the beginning, and verie life of the thing. For unlesse there were composition, the thing should never be brought to passe. Division is a separation of the parts of the compound, & so separation hath bin his conjunction. I tell you againe, that the spirit will not dwell with the body, nor be in it, nor by any meanes abide with it until the body be made subtil & thin as the spirit is. But when it is attenuate and subtill, and hath cast off his thicknes, & put on thinnes, hath forsaken his grossnesse & corpority, & is become spirituall, then shall he be mingled with the subtill spirits, & imbibed in them, so that both shall become one and the same, & they shall not be severed, like as water put to water cannot be divided. Suppose that of two like quantities, that are in solution and congelation, the larger is the soule, the lesser is the body: adde afterward to the quantitie which is the soule, that quantity which is in the body, & it shall participate with the first quantity in vertue only: then worke them as we have

151

wrought them, and so thou shalt obtaine thy desire, and Euclide his line shall be verified unto thee. Afterwarde take his quantity, and know his weight, and give him as much moisture as he will drink, the weight of which moisture we have not here determined. Then againe worke them with an operation unlike the former, first imbibing and subliming it, and this operation is that which they call Albification, and they name it Yarit, that is, Silver, and white Leade. And when thou hast made this compounde white, adde to him so much of the Spirit, as maketh halfe of the whole, and set it to working, till it waxe redde, and then it shall be of the colour Alsulfir, which is verie red, and the Philosophers have likened it to Gold, the effect hereof, leadeth thee to that which Aristotle saide to his Disciple Arda: we call the claye when it is white, Yarit, that is Silver: and when it is red, we name it Temeynch, that is Gold. Whitenesse is that which tincteth Copper, and maketh it Yarit, and that is rednesse, which tincteth Yarit, that is Silver, & maketh it Temeynch, that is Gold. He therefore that is able to dissolue these bodies, to subtiliate them, to make them white and red, and (as I have said) to compound them by imbibing, and convert them to the same, shall without all doubt attaine the masterie, and performe the worke whereof I have spoken unto thee.

CHAP. II.

Of the things and instruments necessarie and fit for this worke.

It behoveth thee to know the vessels in this masterie, to wit Aludela, which the Philosophers have called Church-yards, or Cribbles: because in them the parts are divided, and cleansed, and in them is the matter of the masterie made compleat, perfect, and depured. And every one of these must have a Furnace fit for it, and let either of them have a similitude and figure agreeable to the worke. Mezleme, and many other Philosophers, have named all these things in their bookes, teaching the manner and forme thereof. And thou must know, that herein the Philosophers agree together in their writings, concealing it by signes, and making many books thereof, & instruments which are necessarie in these foure foresaid things. As for the instruments, they are two in number. One is a Cucurbit, with his Alembick: the other is Aludel, that is well made. There are also foure things necessarie to these: that is to say, Bodies, Soules, Spirites, and Waters: of these foure dooth the masterie, and minerall worke consist. These are made plaine in the Philosophers Bookes, I have therefore omitted them in mine, and onely touched those things, which they passed over with silence: which he shall easily

discerne, that is but of indifferent judgement. And
this booke I have not made for the ignorant and
unlearned, but for the wise and prudent.

CHAP. III.

Of the nature of things appertaining to this worke.

Know thou, that the Philosophers have given them
diverse names: for some have called them Mines, some
Animal, some Herball, and some by the name of
Natures, that is Naturall: some other have called
them by certaine other names at their pleasures, as
seemed good unto them. Thou must also know, that
their Medicines are neere to Natures, according as
the Philosophers have said in their bookes, that
Nature commeth nigh to nature, and Nature is like to
nature, and Nature is joined to nature, and Nature
is drowned in nature, and Nature maketh nature
white, & Nature doth make nature red, and generation
is retained with generation, & generation conquereth
with generation.

CHAP. IIII.

Of Decoction, and the effect thereof.

Know thou that the Philosophers have named Decoction in their Bookes, saying, that they make Decoction in thinges: and that is it that engendreth them, and changeth them from their substances and colours, into other substances and colours. If thou transgresse not, I tell thee in this booke, thou shalt proceed rightly. Consider brother, the seed of the earth, whereon men live, how the heate of the Sunne worketh in it, till it be ripe, when men and other creatures seede upon it, and that afterwarde Nature worketh on it by her heate within man, converting it into his flesh and blood. For like hereto is our operation of the masterie: the seed whereof (as the learned have sayde) is such, that his perfection and proceeding consisteth in the fire, which is the cause of his life and death, without somewhat coming betweene, and his spiritualtie, which are not mingled but with the fire. Thus have I tolde thee the truth, as I have seene and done it.

CHAP. V.

Of Subtiliation, Solution, Coagulation, and commistion of the Stone, and of their cause and end.

Know, that except thou subtiliate the bodie till all become water, it will not rust and putrifie, and then it cannot congeale the fitting soules, when the fire toucheth them: for the fire is that which congealeth them by the ayd thereof unto them. And in like manner have the Philosophers commanded to dissolve the bodies, to the end ye heat might enter into their bowels. Again we returne to dissolve those bodies, & congeale them after their solution, with that thing which commeth nigh to it, until we joyne all those things which have beene mingled together, by an apt and fit commixtion, which is a temperate quantitie. Whereupon we joyne fire and water, earth and ayre togither: when the thick hath bin mingled with the thin, & the thinner with the thick, the one abydeth with the other, and their natures are changed and made like, whereas before they were simple, because that part which is generative, bestoweth his vertue upon the subtill, and that is the ayre: for it cleaveth unto his like, and is a part of the generation from whence it receiveth power to move and ascend upward. Cold hath power over the thick, because it hath lost his heate, and the water is gone out of it, and the

thing appeared upon it. And the moisture departed by
ascending, & the subtil part of ye aire, and mingled
itself with it for it is like unto it, and of the
same nature. And when the thicke bodie hath lost his
heat and moisture, and that cold and drinesse hath
power over him, and that their parts have mingled
themselves, and be divided, and that there is no
moisture to joyne the partes divided, the parts
withdraw themselves. And afterwards the part which
is contrary to colde, by reason that it hath
continued, & sent his heat and decoction, to the
parts of ye earth, having power over them, and
exercising such dominion over the cold, that where
before it was in the thicke body, it now lurketh and
lieth hid, his part of generation is changed,
becoming subtil and hot, and striving to dry up by
his heat. But afterward the subtill part (that
causeth natures to ascende) when it hath lost his
accidentall heat, & waxeth cold, then the natures
are changed, and become thicke, and descend to the
center, where ye earthly natures are joined
together, which were subtiliate and converted in
their generation, and imbibed in them: and so the
moisture coupleth together the parts divided: but
the earth endevoureth to drie up that moisture,
compassing it about, and hindering it from going
out: by means wherof, that which before lay hid,
doth now appear: neither can the moisture be

separated, but is retained by the drinesse. And in like manner we see, that whosoever is in the worlde, is retained by or with his contrarie, as heate with colde, and drinesse with moisture. Thus when each of them hath besieged his Companion, the thin is mingled with the thicke, and those things are made one substance: to wit, their soule hote and moyst, and their body colde and drie: then it laboureth to dissolve and subtiliate by his heate and moisture, which is his soule, and striveth to enclose and retaine with his body that is colde and drie. And in this manner, is his office changed and altered from one thing to another. Thus have I tolde thee the truth, which I have both seene & done, giving thee in charge to convert natures from their subtilitie and substances, with heate and moisture, into their substances and colours. Now if thou wouldst proceed aright in this mastery, to obtaine thy desire, passe not the boundes that I have set thee in this booke.

CHAP. VI.

The manner how to fixe the Spirit.

Know also, that when the bodie is mingled with moisture, and that the heate of the fire meeteth therewith, the moisture is converted on the body, and dissolveth it, and then the spirite cannot issue forth, because it is imbibed with the fire. The Spirits are fugitive, so long as the bodies are mingled with them, and strive to resist the fire & his flame: and yet these parts can hardly agree without a good operation and continuall labour: for the nature of the soule is to ascend upward, whereas the center of the soule is. And who is he that is able to joyne two or divers things together, where their centers are divers: unlesse it be after the conversion of their natures, and change of the substance and thing, from his nature, which is difficult to find out? Whosoever therefore can convert the soule into the bodie, the bodie into the soule, and therewith mingle the subtile spirites, shall be able to tinct any body.

CHAP. VII.

Of the Decoction, Contrition, and washing of the stone.

Thou art moreover to understand, that Decoction, contrition, cribation, mundification, and ablution, with sweet waters is very necessary to this secret and mastery: so that he who will bestow any paines herein, must cleanse it very well, and wash the blacknesse from it, and darknes that appeareth in his operation, and subtiliate the bodie as much as he can, and afterwarde mingle therewith the soules dissolved, and spirits cleansed, so long as he thinke good.

CHAP. VIII.

Of the quantitie of the Fire, and of the commoditie and discommoditie of it.

Furthermore, thou must be acquainted with the quantity of the fire, for the benefit and losse of this thing, proceedeth from the benefit of the fire. Wherupon Plato said in his booke: The fire yieldeth profit to that which is perfect, but domoge and corruption to that which is corrupt: so that when his quantitie shall be meete & convenient, it shall prosper, but if it shall exceed measure in things, it shall without measure corrupt both: to wit, the perfect and corrupt: and for this cause it was requisite that the learned should poure their medicines upon Elixir, to hinder and remove from them the burning of the fire, & his heate. Hermes also said to his father. I am afraide Father of the enemie in my house: to whom he made answer, Son take the dog Corascene, & the bitch of Armenia, put them together, and they shall bring a dog of the colour of heaven, and dip him once in the sea water: for he shall keep thy friend, and defend thee from thy enemie, and shall helpe thee whersoever thou become, alwayes abiding with thee, both in this world, and in the world to come. Now Hermes meant by the dog & bitch, such things as preserve bodies from the scorching he ate of the fire. And these things are waters of Calces and Salts, the composition whereof, is to be found in the Philosophers books, that have written of this mastery, among whome, some have named them Sea-waters, and Birdes milke, and such like.

CHAP IX.

Of the Separation of the Elements of the Stone.

Thou must afterward bother, take this precious Stone, which the Philosophers have named, magnified, hidden & concealed, & put it in a Cucurbit with his Alembick, & divide his natures: that is, the foure elements, the Earth, the Water, the Aire, and the Fire. These are the body and soule, the spirit and tincture. When thou hast divided the water from the earth, and the aire from the fire, keep both of them by themselves, and take that which descendeth to the bottom of the glasse, being the lees, and wash it with a warme fire, til his blacknesse be gone, and his thicknesse departed: then make him very white, causing the superfluous moysture to flie away, for then he shall be changed and become a white calx, wherein there is no cloudie darkenesse, nor uncleannesse, and contrarietie. Afterward returne back to the first natures, which ascended from it, and purifie them likewise from uncleanness, blacknesse, and contrarietie: and reiterate these works upon them so often, until they be subtiliate, purified, and made thin: which when thou hast done, thou shalt acknowledge that God hath bin gracious unto thee. Know brother, that this work is one stone, into which Gatib may not enter, that is to say, any strange thing. The learned work with this, and from hence proceedeth a medicine that giveth perfection. There must nothing be mingled herewith, either in part or whole. This Stone is to be found at all times, in every place, and about every man, the search whereof is not troublesome to him that seeketh it, wheresoever he be. This Stone is vile, blacke, and stinking: It costeth nothing: it must be taken alone: it is somewhat heavie, and it is called the Originall of the world, because it riseth up like things that bud forth. This is his revelation and apparance to him that maketh inquirie after it.

CHAP. X.

Of the nature of the Stone, and his birth.

Take it therefore and worke it as the Philosopher hath told you in his booke, when he named it after this manner. Take the Stone, no Stone, or that is not a Stone, neither is of the nature of a Stone. It is a Stone whose myne is in the top of the mountaines: and here by mountaines, the Philosopher understandeth living creatures, wherupon he saide. Sonne, go to the mountaines of India, and to his caves, & pull out thence precious stones which will melt in the water when they are putte into it. And this water is that which is taken from other mountaines and hollow places. They are stones Sonne, and they are not stones, but we call them so for a Similitude which they have to stones. And thou must know, that the rootes of their mines are in the ayre, and their tops in the earth, and it will easily be heard when they are pluckt out of their places, for there will be a great noise. Go with them my sonne, for they will quickly vanish away.

CHAP. XI.

Of the commistion of the Elements that were seperated.

Begin composition, which is the circuite of the whole worke, for there shall be no composition without marriage and putrefaction. The Marriage is to mingle the thinne with the thicke, and Putrefaction, is to rost, grinde, and water, so long till all be mingled together and become one, so that there should be no diversitie in them, nor separation from water mingled with water. Then shall the thicke labour to retaine the thinne: then shall the soule strive with the fire, and endevour to beare it: then shall the Spirite labour to be drowned in the bodyes, and poured forth into them. And this must needes be, because the bodye dissolued, when it is commixt with the Soule, it is likewise commixt with every part therof: & other things enter into other things, according to theyr similitude and likenesse, and are changed into one and the same thing. And for this cause the soule must partake with the commoditie, durablenesse, and permanencie, which the body received in his commixtion. The like also must befall the Spirite in this state or permanencie as the soule and boby: for when the Spirit shall be commixt with the soule by laborious operation, and all his partes with all the

167

partes of the other two, to wit, the soule and bodie, then shall the Spirite and the other two, ee converted into one indivisible thing, according to their entire substance, whose natures have beene preserved, and their partes have agreed and come together: whereby it hath come to passe, that when this compounde hath met with a body dissolved, and that heate hath got hold of it, and that the moisture which was in it appeareth, and is molten in the dissolved body, and hath passed into it, and mixt it selfe with that which was of the nature of moisture, it is inflamed, and the fire defendeth itself with it. Then when the fire would been flamed with it, it will not suffer the fire to take holde of it, that is to say: to cleave unto it with the Spirit mingled with his water. The fire will not abide by it until it be pure. And in like manner doth the water naturally flie from the fire, whereof when the fire hath taken hold, it doth forth with by little and little evaporate. And thus hath the body beene the meanes to retaine the water, and the water to retaine the oyle, that it should not burne nor consume away, and the oyle to retaine tincture, and tincture the precise cause to make the colour appeare and shew forth the tincture, wherein there is neither light nor life. This then is the true life and perfection of the worke and masterie which thou soughtest for. Be wise therefore and

understande, and thou shalt find what thou lookest
for, if it please God.

CHAP. XII.

Of the solution of the Stone compounded.

The Philosophers moreover have taken great paines in dissolving, that the body and soule might the better be incorporate, for all those things that are together in contrition, assation, and rigation, have a certaine affinitie and alliance betweene themselves, so that the fire may spoyle the weaker of nature, till it utterly fade and vanish away, as also it again returneth upon the stronger parts, until the bodie remaine without the Soule. But when they are thus dissolued and congealed, they take the parts one with another, as well great as small, and incorporate them well together, till they be converted and changed into one and the same thing. And when this is done, the fire taketh from the Soule as much as from the body, neither more nor lesse, and this is the cause of perfection. For this cause it is necessary (teaching the composition of Elixir) to afford one chapter for expounding the solution of simple bodyes and soules, because bodyes do not enter into soules, but do rather withhold and hinder them from sublimation, fixation, retention, commistion, and the like operations, except mundification go before. And thou shalt know, that solution is after one of these two wayes: for either it extracteth the inward parts of things unto their

Superficies, and this is solution (an example whereof thou hast in Silver that seemeth cold and drie, but being dissolved, and that his inwards appeare, it is found hot and moyst) or else it is to purchase to a body an accidentall moisture, which it had not before, and to adde hereunto his owne humiditie, whereby his parts may be dissolved, and this likewise is called solution.

CHAP. XIII.

Of the coagulation of the Stone dissolued.

Some among the learned have said, Congeale in a bath with a good congelation as I have tolde thee, and this is Sulphur shining in darknesse, a red Hiasinth, a firy & deadly poyson, the Elixir that abideth upon none, a victorious Lion, a malefactor, a sharpe sworde, a precious Triacle, healing every infirmitie. And Geber the sonne of Hayen sayd, that all the operations of this masterie are contained under fixe things: to put to flight, to melt, to incerate, to make as white as Marble, to dissolve and congeale. That putting to flight, is to drive away and remove blacknesse, from the spirit and soule: the melting is the liquefaction of the body: to incerate belongeth properly to the body, and is the subtiliation thereof: to whiten, is properly to melt speedily: to congeale, is to congeale the body with the soule alreadie prepared. Againe, flight appertaineth to the body and soule: to melt, whiten, incerate, and dissolve, belong unto the body, and congelation to the soule. Be wise and understand.

CHAP XIIII.

That there is but one Stone, and of his nature.

Bauzan, a Greeke Philosopher, when it was demaunded of him, whether a stone may be made of a thing that buddeth, made answere, yea, to wit, the two first stones, the stone Alkali, and our stone, which is the life and workmanship of him that knoweth it: but he that is ignorant of it, and hath not made it, and knoweth not how it is engendred, supposing it to be no stone, or that conceiveth not with himselfe whatsoever I have spoken of it, and yet will make a tryall of it, prepareth himselfe for death, and casteth away his money: for if he cannot find out this precious stone, another shall not arise in his place, neither shall natures triumph over him. His nature is great heate with moderation. He that now knoweth it, hath profited by reading this booke, but he that remaineth ignorant, hath lost his labour. It hath many properties and vertues, for it cureth bodies of their accidentall diseases, and preserveth sound substances, in such sort, that their appeareth in them no perturbations of contraries, nor breach of their bond and union. For this is the sope of bodies, yea their spirit and soule, which when it is incorporate with them, dissolueth them without any losse. This is the life of the dead, and their resurrection, a medicine preserving bodies, and purging superfluities. He that understandeth, let him understand, and he that is ignorant, let him be ignorant still: for it is not to be bought with money, it is neither to be bought nor sold. Conceive his vertue, value, and worth, and then begin to worke: wherof a learned man hath said: God giveth thee not this masterie for thy sole audacity, fortitude & subtilitie, without all labour, but men labor, and God giveth them good successe. Adore then God the creator, that hath vouchsafed thee so great favor in his blessed works.

CHAP. XV.

The manner how to make the Stone white.

Nowe therefore when thou wilt enterpise this worthy worke, thou shalt take the precious stone, and put it in a Cucurbite, covering it with an Alembicke, being well closed with the lute of wisdome, and set it in verie hote dung, then shalt thou distill it, putting a receiver under it, whereinto the water may distill, and thus thou shalt leave it, till all the water be distilled, and moisture dryed up, and that drynesse prevaile ouer it: then shalt thou take it out drie, reserving the water that is distilled, until thou hast neede of it: thou shalt take (I say) the drie bodie that remayned in the bottome of the Cucurbite, and grinde it, and put it in a vessell, in greatnesse answerable to the quantitie of the medicine, and burie it in verie hote horse-dung as thou canst get, the Vessell beeing well shut with the lute of Wisedome, and so let it rest. But when thou perceivest the dung to waxe colde, thou shalt get thee other that is fresh, and very hot, and therein put thy Vessell. Thus shalt thou do by the space of fortie dayes, renuing thy dung so often as occasion shall serue, and the Medicine shall dissolue of itselfe, and become a thicke white water: which when thou beholdest to be so, thou shalt weight it, & put there to so much of the water which thou hast kept, as will make the halfe of his weight, closing thy Vessell with the lute of Wisedome, and put it againe in hote horse-dung, for that is hote and moyst, and thou shalt not omit (as I have sayde) to renue the dung, when it beginneth to coole, till the tearme of fortie dayes be expired: for the Medicine shall be congealed in the like number of dayes, as before it was dissolved in. Again, take it, and note the just weight of it, and according to his quantitie, take of the water which thou madest before, grind the body, and subtiliate it, and poure the water upon it, and set it againe in hot horse-dung, for a weeke and a halfe, that is to say, ten days, then take it out, and thou shalt see that the bodie hath already drunk up the water. Afterward grinde it againe, and put thereto the like quantitie of that water, as thou didst before: bury it in dung, and leave it there for ten dayes more: take it out againe, and thou shalt find that the body hath already drunke up the water. Then (as before) grinde it, putting thereto of the foresayd water, the foresayd quantitie, and bury it in the foresayd dung, and let it rest there ten dayes longer, and afterward draw it out, so shalt thou do the fourth time also: which being done, thou shalt drawe it forth, and grinde it, and burie it in dung till it bee dissolved. Afterward, take it out, and reiterate it yet once more, for then the birth is perfect, and his worke ended. Now when this is done, and that thou hast brought this thing to this honourable estate, thou shalt take two hundred and fiftie drams of Lead, or Steele, and melt it: which being molten, thou shalt cast thereon one dramme of Cinnabarus: that is, of this Medicine, which thou hast brought to this honourable estate, and high degree, and it shall retaine the Steele or Leade, that it fly not from the fire: it shall make it white, and purge it from his drosse and blacknesse, and convert it into a tincture perpetually abiding. Then take a dramme of these two hundred and fiftie, and proiect it upon two hundred and fiftie drammes of Steele or Copper, and it shall convert it into Silver, better then that of the Myne. This is the greatest and last worke that it can effect, if God will.

CHAP. XVI.

The conversion of the foresaid Stone into red.

And if thou desirest to convert this masterie into Gold, take of this medicine (which as I saide, thou hast brought to this honourable estate and excellencie) the weight of one dramme (and this after the manner of thy former example) and put it in a vessell, and bury it in horse-dung for fortie dayes, and it shall be dissolved: then thou shalt give it water of the dissolved body to drink, first as much as amounteth to halfe his weight, afterward until it be congealed, thou shalt bury it in most hot dung, as is above sayd. Then thou shalt orderly proceed in this Chapter of Gold, as thou hast done in the former Chapter of Silver: and it shall be Gold, and make Gold God willing. My Sonne keepe this most secret Booke, and commit it not unto the handes of ignorant men, being a secret of the secretes of God: For by this meanes thou shalt attaine thy desire. Amen.

Here endeth the secrets Alchimy, written in Hebrew by Calid, the son of Iarich.

An excellent discourse of the admirable force and efficacie of Art and Nature, written by the famous Frier Roger Bacon, Sometime fellow of Merton Colledge, and afterward of Brasen-nose in Oxford.

Some there are that aske whether of these twaine be of greatest force, and efficacie, Nature, or Art, whereto I make answere, and say, that although Nature be mightie and maruailous, yet Art using Nature for an instrument, is more powerfull then naturall vertue, as it is to be seene in many thinges. But whatsoeuer is done without the operation of Nature or Art, is either no humane worke, or if it be, it is fraudulently and colourablie performed: for there are some, that by a nimble motion and shewe of members, or through the diversitie of voyces, and subtillitie of instruments, or in the darke, and by consent do propose unto men diuerse things, to be wondred at, that have indeede no truth at all. The world is everywhere full of such fellowes. For Iuglers cogge many things through the swiftnesse of their hands: and others with varietie of voyces, by certaine devices that they have in their bellies, throats or mouthes, will frame mens voyces, farre off, or neare, as it pleaseth them, as if a man spake at the same instant: yea they will counterfeite the soundes of bruit beasts. But the causes hidden in the grasse, or buried in the sides of the earth, prove it to be done by a humane force, and not by a spirit, as they would make men believe. In like manner, whereas they affirm things without life to moue verie swiftly in the twilight of the evening or morning, it is altogether false, and untrue. As for concent, it can faigne any thing that men desire, according as they are disposed togither. In all these neither Physicall reason, nor Art, nor naturall power hath anye place: and for this cause it is more abhominable, sith it contemneth the lawes of Phylosophie, and contrarie to all reason, inuocateth wicked Spirites, that by their help they may have their desire. And herein are they deceived, that they thinke the Spirits to be subject unto them, and that they are compelled at mens pleasures, which is impossible: for humane force is farre inferiour to that of the Spirites. And againe, they fowly erre, to dreame that the cursed spirits are called uppe, and figured, by vertue of those naturall meanes which the thou hast removed the weake from the strong, and put the powder thereto three, foure, or five times, or oftner, always working after one and the selfe same manner. And if thou canst not worke with warme water, thou shalt offer violence. But if it be broken by reason of the tartnesse and tendernesse of the medicine, together with powder thou must verie warily put more Gold to, and mollifie it: but if the plentie of the powder cause it to breake, thou shalt give it more of the medicine, and if it be long of the strength of the water: water it with a Pestill, and gather together the matter so well as thou canst, and separate the water by little and little, and it will returne to his former state. This water thou shalt drie up, for it hath both the powder and water of the medicine, which are to be incorporate as dust. Be not asleepe nowe, for I have tolde thee a great and profitable secret. And if thou couldest tell how to place and sette in order the partes of a burnt shrub, or of a willow, and many such like things, they would naturally keep a union. Beware at any hand that thou forgettest not this, because it is very profitable for many things. Thou shalt mingle the Trinity with the union being first molted, and they will rise up as I suppose like unto the stone Iberus: doubtlesse it is mortified by the vapour of the lead, which lead thou shalt find if thou presse it out of the dead body, and this dead body thou shalt burie in a stillitory. Hold fast this secret, for it is nought worth. And in like manner shalt thou deale with the vapour of a Margarite or the stone Tagus, burying the dead as before thou art commanded.

And now forsooth the yeares of the Arabians being accomplished, I make answere to your demand after this manner. You must have a medicine that will dissolve in a thing that is melted, and be annointed in it, and enter into his second degree, and be incorporate with it, not proving a fugitive servant, and change it, and be mixt with the roote of the Spirit, and bee fixed by the calx of the mettall. Now it is thought that fixation prepareth, when the body & spirit are layd in their place, and sublimed, which must be so often reiterated, til the body be made a spirit, and the spirit become a body. Take therefore of the bones of Adam, and of calx the same weight, there must be sixe for the rocky stone, & five for the stone of unions, & these you must worke together with Aqua vitae (whose property is to dissolve all other things) that it may bee dissolued and boiled in it. And this is a sign of Inceration, if the medicine will melt, when it is poured on an Iron redde hot. This done, poure water into it in a moyst place, or else hang it in the vapour of very hote and liquid Waters, and congeale it in the Sunne then thou shalt take Salt-peter, & convert Argent-vive into lead: and againe, thou shalt wash and grind that therewith, that it may come nigh to Silver, and afterward worke as thou didst before. Moreover, thou shalt drinke uppe all after this sort. Notwithstanding, thou shalt take of Salt-peter, Luru (?) otri, and of Sulphur, and by this meanes make both to thunder and lighten. Thus shalt thou person do the worke. Nowe consider with thy selfe whether speake in a riddle, or tell thee the plaine truth. There be some that have bin of another mind: for it was said unto me, that all things must be resolved to the matter, whereof you may find Aristotle his judgement in vulgar & unknowne places, and therfore I shall hold my peace. Now when thou hast them, thou (?) shall have many simples and equals, and thou shalt effect by contrary things and divers (?), which before I termed the keyes of the (?). And Aristotle saith that the equalitie of the powers doth containe in it the action and passion of bodies and this likewise is the opinion of Auerroes reproving Galen. It is thought that this is the most simple and pure medicine that may be found: It is good against the fevers and passions both of mind and bodyes, more cheape then any medicine whatsoeuer. He that writ these things shall have the key that openeth and no man shutteth, and when he hath shut, no man is able to open it again.

FINIS.

Tract on the

Tincture and Oil of Antimony

by

Roger Bacon
(circa 1220 - 1292)

On the true and right Preparation of Stibium /
to heal human weaknesses and illnesses therewith,
and to improve the imperfect metals.

From Friedrich Roth-Scholtz,

Deutsches theatrum chemicum,

Nürnberg: Adam Jonathan Felsecker, 1731.

Preface

Dear reader, at the end of his Tract on Vitriol,
Roger Bacon mentions that because of the
multiplication of the Tincture that is made from
Vitriol, the lover of Art should acquaint himself
with the Tract De Oleo Stibii. Therefore I
considered that it would be good and useful that the
Tract De Oleo Stibii follows next. And if one
thoroughly ponders and compares these tinctures with
one another, then I have no doubt that one will not
finish without exceptional profit. Yet, every lover
of Art, should mind always to keep one eye on Nature
and the other on Art and manual labour. For, when
these two do not stand together, then it is a lame
work, as when someone thinks he can walk a long path
on one leg only, which is easily seen to be
impossible,
Vale.

Joachim Tanckivs

De Oleo Antimonii Tractatus.

ROGERII BACONIS ANGLI

Summi Philosophi & Chemici.

Stibium, as the Philosophers say, is composed from the noble mineral Sulphur, and they have praised it as the black lead of the Wise. The Arabs in their language, have called it Asinat vel Azinat, the alchemists retain the name Antimonium. It will however lead to the consideration of high Secrets, if we seek and recognize the nature in which the Sun is exalted, as the Magi found that this mineral was attributed by God to the Constellation Aries, which is the first heavenly sign in which the Sun takes its exaltation or elevation to itself. Although such things are thrown to the winds by common people, intelligent people ought to know and pay more attention to the fact that exactly at this point the infinitude of secrets may be partly contemplated with great profit and in part also explored. Many, but these are ignorant and unintelligent, are of the opinion that if they only had Stibium, they would get to it by Calcination, others by Sublimation, several by Reverberation and Extraction, and obtain its great Secret, Oil, and Perfectum Medicinam. But I tell you, that here in this place nothing will help, whether Calcination, Sublimation, Reverberation nor Extraction, so that subsequently a perfect Extraction of metallic virtue that translates the inferior into the superior, may profitably come to pass or be accomplished. For such

189

shall be impossible for you. Do not let yourselves be confused by several of the philosophers who have written of such things, i.e., Geber, Albertus Magnus, Rhasis, Rupecilla, Aristoteles and many more of that kind. And this you should note. Yes, many say, that when one prepares Stibium to a glass, then the evil volatile Sulphur will be gone, and the Oil, which may be prepared from the glass, would be a very fixed oil, and would then truly give an ingress and Medicine of imperfect metals to perfection. These words and opinions are perhaps good and right, but that it should be thus in fact and prove itself, this will not be. For I say to you truly, without any hidden speech; if you were to lose some of the above mentioned Sulphur by the preparation and the burning, as a small fire may easily damage it, so that you have lost the right penetrating spirit, which should make our whole Antimonii corpus into a perfect red oil, so that it also can ascend over the helm with a sweet smell and very beautiful colors and the whole body of this mineral with all its members, without loss of any weight, except for the foecum, shall be an oil and go over the helm. And note also this: How would it be possible for the body to go into an oil, or give off its sweet oil, if it is put into the last essence and degree? For glass is in all things the outermost and least essence. For you shall know that all creatures at

the end of the world, or on the last and coming
judgement of the last day, shall become glass or a
lovely amethyst and this according to the families
of the twelve Patriarchs, as in the families of
jewels which Hermes the Great describes in his book:
As we have elaborately reported and taught in our
book de Cabala.

You shall also know that you shall receive the
perfect noble red oil, which serves for the
translation of metals in vain, if you pour acetum
correctum over the Antimonium and extract the
redness. Yes not even by Reverberation, and even if
its manifold Beautiful colors show themselves, this
will not make any difference and is not the right
way. You may indeed obtain and make an oil out of
it, but it has no perfect force and virtue for
transmutation or translation of the imperfect metals
into perfection itself. This you must certainly
know.

AND NOW WE PROCEED TO THE MANUAL LABOUR, AND THUS THE PRACTICA FOLLOWS.

Take in the Name of God and the Holy Trinity, fine and well cleansed Antimonii ore, which looks nice, white, pure and internally full of yellow rivulets or veins. It may also be full of red and blue colors and veins, which will be the best. Pound and grind to a fine powder and dissolve in a water or Aqua Regis, which will be described below, finely so that the water may conquer it. And note that you should take it out quite soon after the solution so that the water may conquer it. And note that you should take it out quite soon after the solution so that the water will have no time to damage it, since it quickly dissolves the Antimonii Tincture. For in its nature our water is like the ostrich, which by its heat digests and consumes all iron; for given time, the water would consume it and burn it to naught, so that it would only remain as an idle yellow earth, and then it would be quite spoilt.

Consider by comparison Luna, beautiful clean and pure, dissolved in this our water. And let it remain therein for no more than a single night when the water is still strong and full of Spirit.

And I tell you, that your good Luna has then been fundamentally consumed and destroyed and brought to nought in this our water.

And if you want to reduce it to a pure corpus again, then you will not succeed, but it will remain for you as a pale yellow earth, and occasionally it may run together in the shape of a horn or white horseshoe, which may not be brought to a corpus by any art.

Therefore you must remember to take the Antimonium out as soon as possible after the Solution, and precipitate it and wash it after the custom of the alchemists, so that the matter with its perfect oil is not corroded and consumed by the water.

THE WATER; WHEREIN WE DISSOLVE THE ANTIMONIUM, IS MADE THUS:

Take Vitriol one and a half (alii 2. lb.) Sal armoniac one pound, Arinat (alii Alun) one half pound / Sal niter one and a half pound, Sal gemmae (alii Sal commune) one pound, Alumen crudum (alii Entali) one half pound. These are the species that belong to and should be taken for the Water to dissolve the Antimonium.

Take these Species and mix them well among each other, and distill from this a water, at first rather slowly. For the Spiritus go with great force, more than in other strong waters. And beware of its spirits, for they are subtle and harmful in their penetration.

When you now have the dissolved Antimony, clean and well sweetened, and its sharp waters washed out, so that you do not notice any sharpness any more, then put into a clean vial and overpour it with a good distilled vinegar. Then put the vial in Fimum Equinum, or Balneum Mariae, to putrefy forty (al.i four) days and nights, and it will dissolve and be extracted red as blood. Then take it out and examine how much remains to be dissolved, and decant the clear and pure, which will have a red colour, very cautiously into a glass flask. Then pour fresh vinegar onto it, and put it into Digestion as before, so that that which may have remained with the faecibus, it should thus have ample time to become dissolved. Then the faeces may be discarded, for they are no longer useful, except for being scattered over the earth and thrown away. Afterwards pour all the solutions together into a glass retort, put into Balneum Mariae, and distill the sharp vinegar rather a fresh one, since the former would be too weak, and the matter will very quickly become dissolved by the vinegar. Distill it off again, so that the matter remains quite dry. Then take common distilled water and wash away all sharpness, which has remained with the matter from the vinegar, and then dry the matter in the sun, or otherwise by a gentle fire, so that it becomes well dried. It will then be fair to behold, and have a bright red color. The Philosophers, when they have thus prepared our Antimonium in secret, have remarked how its outermost nature and power has collapsed into its interior, and its interior thrown out and has now become an oil that lies hidden in its innermost and depth, well prepared and ready. And henceforth it cannot, unto the last judgement, be brought back to its first essence. And this is true, for it has become so subtle and volatile, that as soon as it senses the power of fire, it flies away as a smoke with all its parts because of its volatility.

Several poor and common Laborers, when they have prepared the Antimonium thus, have taken one part out, to take care of their expenses, so that they may more easily do the rest of the work and complete it, They then mixed it with one part Salmiac, one part Vitro (alii. Nitro, alii. Titro), one part Rebohat, to cleanse the Corpera, and then proceeded to project this mixture onto a pure Lunam. And if the Luna was one Mark, they found two and a half

Loth good gold after separation; sometimes even more. And therewith they had accomplished a work providing for their expenses, so that they might even better expect to attain to the Great Work. And the foolish called this a bringing into the Lunam, but they are mistaken. For such gold is not brought in by the Spiritibus (alii. Speciebus), but any Luna contains two Mark gold to the Loth, some even more. But this gold is united to the Lunar nature to such a degree that it may not be separated from it, neither by Aquafort, nor by common Antimonium, as the goldsmiths know. When however the just mentioned mixture is thrown onto the Lunam in flux, then such a separation takes place that the Luna quite readily gives away her implanted gold either in Aquafort or in Regal, and lets herself separate from it, strikes it to the ground and precipitates it, which would or might otherwise not happen. Therefore it is not a bringing into the Lunam, but a bringing out of the Luna.

But we are coming back to our Proposito and purpose of our work, for we wish to have the Oil, which has only been known and been acquainted with this magistry, and not by the foolish.

When you then have the Antimonium well rubified according to the above given teaching, then you shall take a well rectified Spiritum vini, and pour

196

it over the red powder of Antimony, put it in a gentle Balneum Mariae to dissolve for four days and nights, so that everything becomes well dissolved. If however something should remain behind, you overpour the same with fresh Spiritu vini, and put it into the Balneum Mariae again, as said before, and everything should become well dissolved. And in case there are some more faeces there, but there should be very little, do them away, for they are not useful for anything. The Solutiones put into a glass retort, lute on a helm and connect it to a receiver, also well luted, to receive the Spiritus. Put it into Balneum Mariae. Thereafter you begin, in the Name of God, to distill very leisurely at a gentle heat, until all the Spiritus Vini has come over. You then pour the same Spiritum that you have drawn off, back onto the dry matter, and distill it over again as before. And this pouring on and distilling off again, you continue so often until you see the Spiritum vini ascend and go over the helm in all kinds of colours. Then it is time to follow up with a strong fire, and a noble blood red Oleum will ascend, go through the tube of the helm and drip into the recipient. Truly, this is the most secret way of the Wise to distill the very highly praised oil of Antimonii, and it is a noble, powerful, fragrant oil of great virtue, as you will hear below in the following. But here I wish to teach and instruct you who are poor and without means to expect the Great Work in another manner; not the way the ancients did it by separating the gold from the Luna. Therefore take this oil, one lot, [ancient weight unit used for the weighing of gold and silver coins - about 1/30 pound] eight lot of Saturn calcined according to art, and carefully imbibe the oil, drop by drop, while continuously stirring the calx Saturni. Then put it ten days and nights in the heat, in the furnace of secrets, and let the fire that this furnace contains, increase every other day by one degree. The first two days you give it the first degree of fire, the second two days you give it the second degree, and after four days and nights you put it into the third degree of fire and let it remain there for three days and nights. After these three days you open the window of the fourth degree, for which likewise three days and nights should be sufficient. Then take it out, and the top of the Saturnus becomes very beautiful and of a reddish yellow colour. This should be melted with Venetian Boreas. When this has been done, you will find that the power of our oil has changed it to good gold. Thus you will again have subsistence, so that you may better expect the Great Work. We now come back to our purpose where we left it earlier. Above you have heard, and have been told to distill the Spiritum vini with the Oleum

Antimonii over the helm into the recipient as well as the work of changing the Saturnum into gold. But now we wish to make haste and report about the second tinctural work. Here it will be necessary to separate the Spiritum vini from the oil again, and you shall know that it is done thus:

Take the mixture of oil and wine spirit put it into a retort, put on a helm, connect a receiver and place it all together into the Balneum Mariae. Then distill all the Spiritum vini from the oil, at a very gentle heat, until you are certain that no more Spiritus vini is to be found within this very precious oil. And this will be easy to check; for when you see several drops of Spiritu vini ascend over the helm and fall into the recipient, this is the sign that the Spiritus vini has become separated from the oil. Then remove the fire from the Balneo, though it was very small, so that it may cool all the sooner. Now remove the recipient containing the Spiritu vini, and keep it in a safe place, for it is full of Spiritus which it has extracted from the oil and retained. It also contains admirable virtues, as you will hear hereafter.

But in the Balneo you will find the blessed blood red Oleum Antimonii in the retort, which should be taken out very carefully. The helm must be very slowly removed, taking care to soften and wash off the Lute, so that no dirt falls down into the beautiful red oil and makes it turbid. This oil you must store with all possible precaution so that it receives no damage. For you now have a Heavenly Oil that shines on a dark night and emits light as from a glowing coal. And the reason for this is that its innermost power and soul has become thrown out unto the outermost, and the hidden soul is now revealed and shines through the pure body as a light through a lantern: Just as on Judgement Day our present invisible and internal souls will manifest through our clarified bodies, that in this life are impure and dark, but the soul will then be revealed and seen unto the outermost of the body, and will shine as the bright sun. Thus you now have two separate things: Both the Spirit of Wine full of force and wonder in the arts of the human body: And then the blessed red, noble, heavenly Oleum Antimonii, to translate all diseases of the imperfect metals to the Perfection of gold. And the power of the Spiritual Wine reaches very far and to great heights. For when it is rightly used according to the Art of Medicine: I tell you, you have a heavenly medicine to prevent and to cure all kinds of diseases and ailments of the human body. And its uses are thus, as follows:

AGAINST PODAGRA or GOUT

In the case of gout one should let three drops of
this Spiritu vini, that has received the power of
the Antimony, fall into a small glass of wine. This
has to be taken by the patient on an empty stomach
at the very moment in time when he sense the
beginning or arrival of his trouble, bodily ailment
and pain. On the next day and afterwards on the
third day it should also be taken and used in the
same way. On the first day it takes away all pain,
however great it may be, and prevents swelling. On
the second day it causes a sweat that is very
inconstant, viscous and thick, that smells and
tastes quite sour and offensive, and occurs mostly
where the joints and limbs are attached. On the
third day, regardless of whether any medicine has
been taken, a purging takes place of the veins into
the bowels, without any inconvenience, pain or
grief. And this demonstrates a great power of
Nature.

AGAINST LEPROSY

To begin with the patient is given six drops on an
empty stomach. And arrange it so that the unclean
person is alone without the company of any healthy
people, in a separate and convenient place. For his

whole body will soon begin to smoke and steam with a stinking mist or vapor. And on the second day his skin will start to flake and much uncleanliness will detach itself from his body. He should then have three more drops of the medicine ready, which he should take and use in solitude on the fourth day. Then on the eighth or ninth day, by means of this medicine and through the bestowal of Divine mercy and blessing, he will be completely cleansed and his health restored.

AGAINST APOPLEXIA OR STROKE

In the case of stroke, let a drop of the unadmixed tincture fall onto the tongue of the person in need. At once it will raise itself and distribute itself like a mist or smoke, and rectify and dissolve the struck part. But if the stroke has hit the body or other members, he should be given three drops at the same time in a glass of good wine, as previously taught in the case of Podagra.

AGAINST HYDROPE OR DROPSY

In the case of dropsy give one drop each day for six days in a row, in Aqua Melissae or Valerianae. On the seventh day give three drops in good wine. Then it is enough.

AGAINST EPILEPSIA, CATALEPSIA, & ANALEPSIA.

In case of the falling sickness, give him two drops at the beginning of the Paroxismi in Aqua Salviae, and after three hours again two drops. This will suffice. But if further symptoms should occur, then give him two more drops as above.

AGAINST HECTIC

In case of consumption and dehydration, give him two drops in Aqua Violarum the first day. On the second day, give him two more drops in good wine.

AGAINST FEVER

In cases of all kinds of hot fevers, give him three drops in a well distilled St. Johnswort water or Cichorii at the beginning of the Paroxismi. Early in the morning on the following day, again give him three drops in good wine on an empty stomach.

AGAINST PEST

In the case of pestilence give the patient seven drops in a good wine, and see to it that the infected person is all by himself, and caused to

sweat. Then this poison will, with Divine
assistance, do him no harm.

FOR THE PROLONGATION AND MAINTENANCE OF A HEALTHY LIFE.

Take and give at the beginning and entry of spring,
when the sun has entered the sign of Aries, two
drops; and at the beginning with God's help, be safe
and protected against bad health and poisoned air,
unless the incurred disease was predestined and
fatally imposed upon man by the Almighty God.

But we now wish to proceed to the Oleum Antimonii
and its Power, and show how this oil may also help
the diseased and imperfect metallic bodies. Take in
the Name of God, very pure refined gold, as much as
you want and think will suffice. Dissolve it in a
rectified Wine, prepared the way one usually makes
Aquam Vitae. And after the gold has become
dissolved, let it digest for a month. Then put it
into a Balneum, and distill off the spiritum vini
very slowly and gently. Repeat this several times,
as long and as often until you see that your gold
remains behind in fundo as a sap. And such is the
manner and opinion of several of the ancients on how
this oil may also help the diseased and imperfect
metallic bodies.

Take, in the Name of God, very pure refined gold, as much as you want and think will suffice. Dissolve it in a rectified Wine, prepared the way one usually makes Aquam Vitae. And after the gold has become dissolved, let it digest for a month. Then put it into a Balneum, and distill off the spiritum vini very slowly and gently. Repeat this several times, as long and as often until you see that your gold remains behind in fundo as a sap. And such is the manner and opinion of several of the ancients on how to prepare the gold. But I will show and teach you a much shorter, better and more useful way. Viz. that you instead of such prepared gold take one part Mercurii Solis, the preparation of which I have already taught in another place by its proper process. Draw off its airy water so that it becomes a subtle dust and calx. Then take two parts of our blessed oil, and pour the oil very slowly, drop by drop onto the dust of the Mercurii Solis, until everything has become absorbed. Put it in a vial, well sealed, into a heat of the first degree of the oven of secrets, and let it remain there for ten days and nights. You will then see your powder and oil quite dry, such that it has become a single piece of dust of a blackish grey colour. After ten days give it the second degree of heat, and the grey and black colour will slowly change into a whiteness so that it becomes more or less white. And at the end of these ten days, the matter will take on a beautiful rose white. But this may be ignored. For this colour is only due to the Mercurio Solis, that has swallowed up our blessed oil, and now covers it with the innermost part of its body. But by the power of the fire, our oil will again subdue such Mercurium Solis, and throw it into its innermost. And the oil with its very bright red colour will rule over it and remain on the outside. Therefore it is time, when twenty years (sic) have passed, that you open the window of the third degree [The alchemical ovens had small openings at different heights, by means of which the heat was regulated.] The external white colour and force will then completely recede inwardly, and the internal red colour will, by the force of the fire, become external. Keep also this degree of fire for ten days, without increase or decrease. You will then see your powder, that was previously white, now become very red. But for the time being this redness may be ignored (is of no consequence), for it is still unfixed and volatile; and at the end of these ten days, when the thirtieth day has passed, you should open the last window of the fourth degree of fire, Let it stay in this degree for another ten days, and this very bright red powder will begin to melt. Let it stay in flux for these ten days. And when you take it out you will find on the bottom a very bright red and transparent stone, ruby colored, melted into the shape of the vial. This stone may be used for Projection, as has been taught in the tract on Vitriol. Praise God in Eternity for this His high revelation, and thank Him in Eternity. Amen.

ON THE MULTIPLICATION LAPIDIS STIBII.

The ancient sages, after they had discovered this stone and prepared it to perfect power and translation of the imperfect metals to gold, long sought to discover a way to increase the power and efficiency of this stone. And they found two ways to multiply it: One is a multiplication of its power, such that the stone may be brought much further in its power of Transmutation. And this multiplication is very subtle, the description of which may be found in the Tract on Gold. The second multiplication is an Augmentum quantitatis of the stone with its former power, in such a way that it neither loses any of its power, nor gains any, but in such a manner that its weight increases and keeps on increasing ever more, so that a single ounce grows and increases to many ounces. To achieve this increase or Multiplication one has to proceed in the following manner: Take in the Name of God, your stone, and grind it to a subtle powder, and add as much Mercurii Solis as was taught before. Put these together into a round vial, seal with sigillo Hermetis, and put it into the former oven exactly as taught, except that the time has to be shorter and less now. For where you previously used ten (alii thirty) days, you may now not use more than four (alii ten) days. In other respects the work is exactly the same as before. Praise and thank God the Almighty for His high revelation, and diligently continue your prayers fir His Almighty Mercy and Divine blessings of this Work and Art as well as His granting you a good health and fortuitous welfare. And moreover, take care always to help and counsel the poor.

LAVS DEO OMNIPOTENTI

-Finis-

Frier BACON

HIS

DISCOVERY

OF THE

MIRACLES OF ART, NATURE AND MAGICK.

Faithfully translated out of Dr. Dees

Own Copy, by T.M. and never

Before in English.

LONDON

Printed for Simon Miller at the Starre

In St. Pauls Church-yard, 1659.

This edition is from a copy in the British Museum.

The Translator to the Reader.

A Prejudicate eye much lessens the noblenesse of the Subject. Bacon's name may bring at the first an inconvenience to the Book, but Bacon's ingenuity will recompence it ere he be solidly read. This as an Apology is the usher to his other Workes, which may happily breath a more free Air hereafter, when once the World sees how clear he was, from loving Necromancy. 'Twas the Pope's smoak which made the eyes of that Age so sore, as they could not discern any open hearted and clear headed Soul from an heretical Phantasme. The silly Fryers envying his too prying head, by their craft had almost got it off his shoulders. It's dangerous to be wiser than the multitude, for that unruly Beast will have every over-topping head to be lopped shorter, lest it plot, ruine, or stop the light, or shadow its extravagancies. How famous this Frier is in the judgement of both godly and wise men, I referre you to the Probatums of such men, whose single Authorities were of sufficiency to equalize a Jury of others; and as for the Book, I refer it to thy reading. As for myself, I refer me to him, whom I serve, and hope thou wilt adore.

The Judgement of Divers Learned Men

Concerning Fryer Bacon.

IO Selden de Diis Syris Sintag.I. r.2.---7.25.

That singular Mathematician, learned beyond what the Age he liv'd in did ordinarily bring forth, Roger Bacon an Oxford man, and a Fryer minorite.

The Testimony of Gabriel Powel in his Book of Antichrist, p.14.

Roger Bacon an Englishman, a founded Scholar of Merton-College in Oxford, a very quick Philosopher, and withal a very famous Divine, he had an incredible knowledge in the Mathematicks, but without Necromancy (as John Balleus doth report) although he be defamed for it by many: Now this man after he had sharply reproved the times wherein he liv'd; these Errours, saith he, speak Antichrist present. Nicholas the Fourth Pope of Rome did condemn his Doctrine in many things, and he was by him kept in prison for many years together; as Antonine hath it in his Chronicle. He flourished in the year of our Lord, 1270.

John Gerhard Vossius in his Book of the four Popular Arts, printed at Amsterdam, 1650. Is everywhere full of the praises of Bacon, as in the year 1252. About these mens time Roger Bacon also flourished, an Englishman, and a Monk of the Order of St. Francis, who as he had div'd into all Arts and Sciences: so also he writ many things of them, he was a man most learned and subtil unto a Miracle, and did such wonderful things by the help of Mathematicks, that by such as were envious and ignorant, he was accused of Diabolical Magick, before Pope Clement the 4th, and for that cause was detained in prison by him for some time. Jo. Pecus Earl of Mirandula, the Phenix of all the wits of his Age, calls him likewise, very ingenious. Moranlicus also commends highly his Opticks. He was buried at Oxford in the Monastery of the Monks of his own Order, anno. 1255. So Chap. 35. S.32 anno 1255. So Chap. 60. S. 13. Of Musick, anno 1270. So Chap. 70. S. 7., 1270. Roger Bacon a Franciscan Monk, and a Divine of Oxford, was famous amongst the English in all sorts of Sciences, a man of so vast learning, that neither England, no nor the world beside, had almost any thing like or equal to him. And either by envy or ignorance of the Age, wherein he lived, was accused of Magick. He in the mean time did write and recommend to the memory of Posterity, a Book of Weights, of the Centers of heavy things, of the Practicks of Natural Magick, &c. For he was a man well vers'd in all sorts of study, very learned in the Latine, Greek and Hebrew

Tongues, a Mathematician every way accomplish, and very skilfull both in Philsophy, Physick, Law and Divinity.

A

LETTER

SENT BY

Frier ROGER BACON

TO

William of Paris,

Concerning both

The Secret Operation

OF

NATURE & ART,

As also

The Nullity of Magick.

Chap. I.

Of and against ficticious Appearances and Invocation of Spirits.

That I may carefully render you an answer to your desire, understand, Nature is potent and admirable in her working, yet Art using the advantage of nature as an instrument (experience tells us) is of greater efficacy than any natural activity.

Whatsoever Acts otherwise than by natural or artificial means, is not humane, but merely ficticious and deceitfull.

We have many men that by the nimblenesse and activity of body, diversification of sounds, exactness of instruments, darkness, or consent, make things seem to be present, which never were really existent in the course of Nature. The world, as any judicious eye may see, groans under such bastard burdens. Jugler by an handsome sleight of hand, will put a compleat lie upon the very sight. The Pythonisse sometimes speaking from their bellies, otherwhile from the throat, than by the mouth, do create what voices they please, either speaking at hand, or farre off, in such a manner, as if a Spirit discoursed with a man, and sometimes as though

Beasts bellowed, which is all easily discovered by private laying hollow Canes in the grasse, or secret places, for so the voices of men will be known from other creatures.

When inanimate things are violently moved, either in the Morning or Evening twilight, expect no truth therein, but down-right cheating and cousenage.

As for consent, men by it may undertake any thing they please, if so be they have a mutual disposition.

These I mention, as practices wherein neither philosophical Reasons, Art, or power of Nature is prevalent. Beyond these there is a more damnable practice, when men despising the Rules of Philosophy, irrationally call up wicked Spirits, supposing them of Energy to satisfie their desires. In which there is a very vast errour, because such persons imagine they have some authority over Spirits, and that Spirits may be compelled by humane authority, which is altogether impossible, since humane Energy or Authority is inferiour by much to that of Spirits. Besides, they admit a more vast mistake, supposing such natural instruments, as they use, to be able either to call up, or drive away any wicked Spirit. And they continue their mistake in endeavouring by Invocations, Deprecations or Sacrifices to please Spirits, making them propitious to their design. Without all question, the way is incomparably more easie to obtain anything, that is truly good for men, of God, or good Angels, then of wicked Spirits. As for things which are incommodious for men, wicked Spirits can no further yield assistance, then they have permission, for the sins of the sonnes of men, from that God, who governs and directs all humane affairs. Hence therefore I shall conclude (these things being beyond, or rather against the Rules of Wisdome) No true Philosopher did ever regard to work by any of these six ways.

Chap. II.

Of Charms, Figures, and their Use.

What men ought to believe touching Figures, Charms, and such stuff, I shall deliver my opinion. Without doubt there is nothing in these dayes of this kind, but what is either deceitfull, dubious, or irrational, which Philosophers formerly invented to hide their secret operations of Nature and Art from the eyes of an unworthy generation. For instance, if the virtue of the Load-stone, whereby it draws iron to it were not discovered, some one or other who hath a mind hereby to cosen the people, so goes about his businesse, as lest any by-stander should discover the work of attraction to be natural, he casts Figures, and mutters forth some Charmes. Thus many things lie dark in Philosophical writings; in which the wiser sort of Readers will expresse so much discretion, as reject the Figures and Charmes, eyeing the works of Nature and Art, that so they may see the mutual concurrence of animate and inanimate creatures, occasioned by Natures conformity, not any efficacy of Figures or Charmes. This is the cause why the unlearned crew have judged such natural or artificial operations to be meerly Magical. And some fond Magicians believe, That their casting of Figures and Charmes was the sole cause of such operations; hereupon leaving their natural and artificial operations have stuck close to their erroneous casting of Figures and Charms. And thus they both have by their own folly deprived themselves of the benefit of the others wisdome. In times past, godly and religious men[1], or rather God himself, or his good Angels composed several Prayers, which yet may retain their primitive virtue. As to this day, in several Countreys, certain prayers are made over hot irons, and water in the River, &c. By which the innocent are cleared, and guilty condemned; yet all this is done by the Authority of the Church, and her Prelates. Our Priests exercise their holy water, as formerly the Iews did in the Old Testament[2], in making the water of Tryal, whereby the wife was tryed, whether she were an adultresse, or honest. Not to instance in others of the like nature. Concerning those Secrets, which are revealed in Magicians writings, although they may contain some truth, yet in regard those very truths are enveloped with such a number of deceits, as it's not very easie to judge betwixt the truth and falsehood, they ought all worthily to be rejected. Neither must men be believed, who would assure us, That Solomon, or some other of our sage Progenitors were Authors of such Books, because those Books are not received either by the Churches

---

[1] Numb.6.27.
[2] Numb.5.

Authority, or by any prudent men, but only by a few cheating Companions to be the works of such men. Mine own experience assures me they compose and set forth new works and inventions of their own, in lofty high flown expressions, the more colourably to make their lies passe under the shelter of the Text; prefixing some specious titles, the better to set them off, impudently ascribe such bastard births to famous Authors.

Figures are either composed of words involved in the formes of letters, invented to contain the sense of some speech[3] or prayer; or they are made according to the face of the Heavens in proper and select seasons. The Figures of the former sort must have the same sentence that I gave of prayers formerly; as for Figures and Impressions[4] of the other kind, unlesse they be made in their peculiar seasons, they are not of any efficacy. And hence it is that all wise men think they effect nothing, who only go according to their prescribed Characters, not at all regarding more than the bare external forme. The more knowing sonnes of Art, dispose all their works of Nature and Art according to the power of the Heavens, casting their work under a right Constellation, no lesse than the casting it in a

---

[3] Oraionis
[4] Sigillis

right Figure. Now in regard there is much difficulty to discern the motion of Celestial Bodies, many are cousened, and very few know, how to begin their work either profitably or truly. Hence it comes to passe, that the croud of judicious Mathematicians and Starre gazers effect little, and that unprofitable, while the more expert Professours, who sufficiently understand their own Art, attain many conveniences both by their Operations and Judgements in select and proper opportunities: And yet let us take notice, how the Physician, or he that would re-erect a drooping soul, effects his designe by the use of Figures or Charmes, which in themselves are meerly fictitious (as Constantine the Physician is of opinion.) Physicians use Figures or Charmes,[5] not for any prevalency in them, but that the raising of the soul is of great efficacy in the curing of the body, and raising it from infirmity to health, by oy and confidence is done by Charmes; for they make the Patient receive the Medicine with greater confidence and desire, exciting courage, more liberal belief, hope and pleasure. The Physician then who would magnifie his cure, may work some way of exciting hope and confidence in his Patient; not that hereby he should cheat, but stirre up the sick to believe he shall recover, which if we pin our faith on

---

[5] Thus some think the Kings evil is cured by creating a belief the reach of the King can cure.

Constantines sleeve[6], is very tolerable. Upon this account he defends the hanging Charmes or Figures about the Neck. The soul no question is of much prevalency by reason of its strong affections over its proper body, as Avicen saith in lib. de anima, & 8. & animal. to which all wise men accord. Hereupon it was, that they concluded sick persons should be delighted by the company of children to play before them, and other pleasing objects. Yea they frequently consent to such things as please the appetite, though they be obnoxious to their disease; because affection, desire and hope of the soul conquers many diseases.

---

[6] This may be done lawfully, if the party that is the principal agent doth nothing by way of compact with any Spirit, or sinistrously.

Chap. III.

SERMONIS.

Of the force of Speech, and a Check to Magick.

In regard truth must not receive the least injury, we should take more exact notice how every agent communicateth the Virtue and Species which is in it to other extrinsecal objects; I mean not only the substantial Virtue, but even Active Accidents, such as are in tertia specie Qualitatis.

As for the Virtues which flows from the Creature, some of them are sensible, some insensible. Man which is both the most noble corporeity, and dignified rational soul, hath no lesse than other things heat and spirits exhaling from him, and so may no lesse than other things emit and dispose of his Virtues and Species to external Objects.

Some creatures we know have power to metamorphose and alter their objects. As the Basilisk,[7] who kills by sight alone. The Wolf, if she first see a man before the man see him, makes the man hoarse.[8] The Hyaena suffers not the dog which comes within his shadow to bark (as Solinus de mirabilious mundi, and

---

[7] Plin. Nat. Hist. lib.39.c.4.
[8] Plin. Lib. 8. Cap. 22. Solin. Poly. Cap 8. Plin. Lib.8.cap.30.Solin.c 30.

others). And Aristotle lib. 2. de Vegetab. saith, That Female Palm-trees bring forth fruit to maturity by the smell of their Males. And Mares in some Kingdoms impregnate by the smell of Horses[9] (as Solinus affirms.) Aristotle in his Secrets assures us of several other contingencies which issue from the Species and Virtues of Plants and Animals. Hence I argue, If Plants and Animals, which are inferiour in dignity to our humane Nature, can emit, then surely may man more abundantly emit Species, Virtues and Colours to the alteration of external Bodies. To this purpose is that, which Aristotle tels us (Lib. de somne & Vigiliâ) a menstruous woman looking in a glasse, doth infect it with spots,[10] like clouds of bloud. Solinus further writes, That in Scythia there are women which have two sights in one eye. (Hence Ovid, Nocet pupilla duplex.) and that these women by their glances kill men. And we our selves know,[11] That men of an evil complexion, full of contagious infirmities, as Leprosie, the Falling-sickness, spotted Feaver, bleer-eyed, or the like, infects those men in their company: While on the other side, men of a sound and wholsome complexion, especially young men, do by their very presence exhilerate and comfort others; which no question,[12] as Galen in his

---

[9] Cap.30. ex C. Plin. Olib.4.c.32. & lib.8.c.42.
[10] Cap.6. vid. & C.Plin. 1.7.c.2.
[11] In the Northern Country some are said to have an evil eye, and to do harm by their looks, yea though they do it not voluntarily.

Techne, proceeds from their pure spirits wholsome and delightsome vapours, their sweet natural colour, and from such Species and Virtues as they emit.

That man whose soul is defiled with many hainous sins, his Body infirme, his Complexion evil, and hath a vehement fancy and desire to hurt his neighbour, may bring more inconveniencies, then another man. The Reason may be, the Nature of Complexion and infirmity yeelds obedience to the thoughts of the Heart, and is more augmented by the intervention of our desires. Hence it is that a leprous person, who is solicitous, desirous and fancying to infect some one or other in the room, may more easily and forceably effect it, than he which hath no such intention, fancy or desire. For (as Avicen observes in the fore-cited place) the nature of the body is obedient to the thoughts, and more intent fancies of the soul. And (as Avicen in the 3ᵈ Metaph. affirms) the thought is the first mover, after that the desire is made conformable to the thought, then after that the natural virtue, which is in the members, obeys the desire and thought; and thus it is both in good and bad effects. Hence it is that a young man of a good Complexion, healthfull, fair, well featured Body,

---

[12] The soul sinful or not, works morally, not physically to the hurt of others, but the man who hath a body may do something Medicante corpore.

having his soul not debauched with sinne, but of a strong fancy and vehement desire to compasse the effecting of some magnificent designe, withall adding the power of his Virtues, Species and natural heat; He may by the force of these Spirits,[13] Vapours and influences work both more powerfully and vehemently, than if he should want any of these fore-going qualifications, especially strong affections and forceable imaginations. Hence I conclude, Men by the concurrence of the foresaid Causes, Words and Works being the Instruments, bring great undertakings to perfection.

As for words, they are hatched within, by the thoughts and desires of the mind, sent abroad by heat, Vocale arteries, and motion of the Spirits. The places of their generation are in open passages, by which there is a great efflux of such spirits, heat, vapours, virtues, and Species, as are made by the soul and heart. And therefore words may so farre cause alterations by these parts or passages, as their Nature will extend. For it's evident, That breathings, yawnings, several resolutions of Spirits and heat come thorow these open passages from the heart and inward parts: Now if these words come from an infirm and evil complexionated body, they are constantly obnoxious. But if from a pure sound and

---

[13] Al. Species.

wholsome constitution, they are very beneficial and comfortable. It's clear then, That the bare generation and prolation of words joyned with desire and intention are considerable in natural operations. Hereupon we do justly say, Vox viva magnam habet virtutem; Living words are of great Virtue. Not that they have any such Virtue of doing or undoing, as Magicians speak of, but only they have the Virtue of Nature, which makes me put in this Caution of being extream cautelous herein. For a man may, as many have already done, erre on both hands: Some wholly denying any operation of words: Others superfluously decline to a Magical use thereof. Our duties should be to have a care of such Books, as are fraught with Charms, Figures, Orizons, Conjurations, Sacrifices, or the like, because they are purely Magical. For instance, the Book De Officiis Spirituum, liber de morte animae, liber de art notariâ, with infinite others, containing neither precepts of Nature or Art, having nothing save Magical Fopperies. Yet herewithall we must remember, there are many Books commonly reputed to be Magical, but have no other fault then discovering the dignity of wisdome. What Books are suspicious, and what not; Every discreet Readers experience will show him. The Book which discovers natural or artificial operations imbrace; that which is void of either or leave both, as suspitious and unworthy the consideration of any wise man. 'Tis usual with Magicians, to treat of both unnecessary and superfluous subjects. 'Twas excellently said of Isaac (in lib. de Febribus), The rational soul is not impeded in its operations, unlesse by the Manicles of ignorance. And Aristotle is of opinion, (in lib. secret.) That a clear and strong intellect, being impregnated by the influences of divine Virtue, may attain to any thing which is necessary. And in 3^d Meteor, he saith, There is no influence or power, but from God. In the Conclusion of his Ethicks, There is no Virtue, whether Moral or Natural without divine influence. Hence it is, that when we discourse of particular agents, we exclude not the Regiment of the universal Agent, and first Cause of all things. For every first Cause hath more influence on the Effect, than any second Cause, as he speaks in the first proposition of Causes.

Chap. IV.
Of admirable Artificial Instruments.

That I may the better demonstrate the inferiority
and indignity of Magical power to that of Nature or
Art, I shall a while discourse on such admirable
operations of Art and Nature, as have not the least
Magick in them, afterwards assign them their Causes
and Frames. And first of such Engines, as are purely
artificial.

It's possible to make Engines to sail withall, as
that either fresh or salt water vessels may be
guided by the help of one man, and made sail with a
greater swiftness, than others will which are full
of men to help them.

It's possible to make a Chariot move with an
inestimable swiftnesse (such as the Currus falcati
were, wherein our fore fathers of old fought,) and
this motion to be without the help of any living
creature.

It's possible to make Engines for flying, a man
sitting in the midst whereof, by turning onely about
an Instrument, which moves artificiall Wings made to
beat the Aire, much after the fashion of a Birds
flight.

It's possible to invent an Engine of a little bulk, yet of great efficacy, either to the depressing or elevation of the very greatest weight, which would be of much consequence in several Accidents: For hereby a man may either ascend or descend any walls, delivering himself or comrads from prison; and this Engine is only three fingers high, and four broad.

A man may easily make an Instrument, whereby one man may in despight of all opposition, draw a thousand men to himself, or any other thing, which is tractable.

A man may make an Engine, whereby without any corporal danger, he may walk in the bottome of the Sea, or other water. These Alexander (as the Heathen Astronomer assures us) used to see the secrets of the deeps.

Such Engines as these were of old, and are made even in our dayes. These all of them (excepting only that instrument of flying, which I never saw or know any, who hath seen it, though I am exceedingly acquainted with a very prudent man, who hath invented the whole Artifice) with infinite such like inventions, Engines and devices are feasable, as making of Bridges over Rivers without pillars or supporters.

Chap. V.

Of Perspective Artificial Experiences.

The physical figuration of rayes are found out to be
very admirable. Glasses and Perspectives may be
framed, to make one thing appear many, one man an
Army, the Sun and Moon to be as many as we please.
As Pliny in the 2^d Book, Nat. Hist. chap. 30 saith,
That Nature so disposeth of vapours, as two Sunnes,
and two Moons; yea sometimes three Sunnes shine
together in the Air. And by the same Reason one
thing may in appearance be multiplied to an
infinity, in regard that after any creature hath
exceeded his own virtue (as Aristotle cap. de
vacuo.) no certain bounds is to be assigned it.

This designe may seem advantagious to strike
terrours into an Enemies Camp or Garison, there
being a multiplication of appearances of Srarres, or
men assembled purposely to destroy them: Especially
if the following designe be conjoyned to the former
(viz.) Glasses so cast, that things at hand may
appear at distance, and things at distance, as hard
at hand: yea so farre may the designe be driven, as
the least letters may be read, and things reckoned
at an incredible distance, yea starres shine in what
place you please. A way, as is verily believed,
Julius Caesar took by grear Glasses from the Coasts

of France, to view the site and disposition of stoth the Castles and Sea-Towns in Great Britain. By the framing of Glasses, bodies of the largest bulk, may in appearance be contracted to a minute volumne, things little in themselves show great, while others tall and lofty appear low and creeping, things creeping and low, high and mighty, things private and hidden to be clear and manifest. For as Socrates did discover a Dragon, whose pestiferous breathings and influences corrupted both City and Countrey thereabouts, to have his residence in the Caverns of the Mountains. So may any other thing done in an Enemies Camp or Garison, be discovered. Glasses may be framed to send forth Species, and poisonous infectious influences, whither a man pleaseth. And this invention Aristotle shewed Alexander, by which he erecting the poison of a Basilisk upon the Wall of a City, which held out against his Army, conveyed the very poison into the City it self. Glasses may be so framed and placed, as that any man coming into a room, shall undoubtedly imagine he sees heaps of gold, silver, prceious stones, or what you please, though upon his approach to the place he shall perceive his mistake.

It's then folly to seek the effecting that by Magical Illusions, which the power of Philosophy can demonstrate.

To speak of the more sublimate powers of
Figurations, leading and congregating rayes by
several Fractions and reflexions to what distance we
please, so as any object may prove combustible. It's
evident by Perspectives they burn backward and
forward, which Authours have treated on in their
Books. That which is the most strange of Figurations
and Mouldings, is the description of Celestial
Bodies, both according to their Longitude and
Latitude, in such Corporeal Figures, as they
naturally move by their diurnal motion. An Invention
of more satisfaction to a discreet head, than a
Kings Crown.

But this will suffice as to Figurations, though we
might produce infinite prodigies of the like Nature.

Chap. VI.

Concerning Strange Experiments.

TO our former discourse we may adjoyn such works as
are effected without Figurations. We may have an
artificial composition of Saltpeter, and other
ingredients;[14] or of the oil of Red Petrolei, and
other things, or with Maltha, Naphtha, with such
like, which will burn at what distance we please,
with which Pliny reports, Lib. 2. Chap. 104. that he
kept a City against the whole Roman Army: For by
casting down Maltha he could burn a Souldier, though
he had on his Armour. In the next place, to these we
may place the Grecian fire,[15] and other combustibles.
To proceed, Lamps may be made to burn, and waters to
keep hot perpetually. For I know many things which
are not consumed in the fire, as the Salamanders
skin Talk, with others, which by some adjunct both
are inflamed and shine, yet are not consumed, but
rather purified. Besides these, we may speak of
divers admirable peeces of Nature.[16] As the making
Thunder and Lightning in the Air; yea with a greater
advantage of horrour, then those which are onely
produced by Nature. For a very competent quantity of
matter rightly prepared (the bignesse of ones thumb)
will make a most hideous noise and corruscation,

---

[14] Oleum rubrum Petroleum.
[15] Ignis Gracus.
[16] Art it shold be, as I suppose.

this may be done several wayes; by which a City or Army may be overcome, much after the fashion as Gideon overcame that vast Army of the Midianites with three hundred men, by the breaking of their Pitchers, and shining of their Lamps, together with the sudden leaping forth of the fire, and inestimable cracklings. These would appear strange, if they were designed to their just height both of proportion and matter. I might produce many strange works of another kind, which though they bring no sensible profit, yet contain an ineffible spectacle of wit, and may be applied to the probation of all such secrets, as the ignorant crew will not imbrace. Such might I name the attraction of Iron to the Loadstone, a thing so incredulous, as none save an eye-witnesse would believe. And in this attraction of Iron, experience will show a diligent searcher, more wonders than any vulgar capacity can entertain.

But to proceed to greater, and more than these. There is an attraction of gold, silver, and all other metals,[17] by a certain stone, much after the same manner. Besides one stone will runne to the heap. Plants may have their mutual concurrence, and the parts of sensible creatures locally divided, will naturally move to a mutual imbracement. The

---

[17] Silver and all other metals. Plin. Hist. 1.36.cap.20. Aliter Vinegar.

consideration whereof makes me think, that there is not any thing, whether in divine or outward matters too difficult for my faith. To proceed higher. The whole power of the Mathematicks may compose a spherical Engine, according to Ptolomies frame in eight Almagest; which sincerely describes both longitude and latitude of all Celestial Bodies; but to give them a natural diurnal motion is not in the power of the Mathematicks. However a discreet head-piece would do well to try the making hereof of such materials and artifice, as it might have a natural diurnal motion. Which seems to me possible; and because many things are moved with the motion of the Heavens, as Comets, the Sea tides, with several other things, which are turned about either in the whole or in part. Such a work might be thought more miraculous, and of a vaster benefit than any thing hitherto mentioned. For the perfecting of this would frustrate all other, whether the more curious, or the more vulgar Astronomical Instruments, which surely would be more valuable than a Kings Coffers; and yet there may matters be brought to passe, which though they will not reach so near a miracle, yet of farre greater publick and private profit. As the producing so much gold or silver, as we please, not by the work of Nature yet accomplishment of Art:[18]

---

[18] Quid sint decem & septem modi auri, octo scilicet ex admixtione argenti cum auro, & primus modis sit.

seeing there may be ten and seven wayes of gold, eight by the mixture of silver with gold; and the first way is made by sixteen parts of gold with some parts of silver, which will attain the four and twentieth degree of gold, always augmenting one degree of gold with one of silver, and so for the mixture of brasse with gold. So the last way is[19] by the four and twenty degrees of pure gold without mixture of other metal. And beyond this, Nature knows no further progresse, as experience tels us. Though Art may augment gold in the degrees of purity, even to infinitenesse, and compleat silver, without the least cheat: And yet that which seems more rare than all this is, That though the rational soul (hath so farre its free-will, as) it cannot be compelled, yet may effectually be excited, induced and disposed freely to alter its affections, desires and behaviours to the dictates of another man. And this may not only be practised upon one particular person, but upon a whole Army, City, or Body of a Nation living under one Region, if we believe experience. And this experience, Aristotle discloseth in his Book of Secrets, both of an Army, Region and single person. And thus I have well nigh finished my thoughts of Nature and Art.

---

[19] Ex from.

Chap. VII.

Of Retarding the Accidents of Old age, and Prolongation of Life.

The furthest attainment, which the complement of Art, joyned with the whole Energy of Nature can reach unto, is the Prolongation of Life to a very old date. How farre this is attainable, manifold experience hath shewed us. Pliny reports, That Pollio, a man of a strong body and mind,[20] lived much longer then men usually now: of whom Octavius Augustus enquiring, What course he took to live so long? was answered aenigmatically, he used Oyl without, and Mulsum within (now according to the opinion of some, it's eight parts of water, and nine of honey) I might produce many examples of the same quality: as that which fell out in the dayes of King William; A Countreyman plowing in the field, found a golden vessel, containing a certain liquor, which he supposing to be the Dew of Heaven, washed his face withall, and drunk of it, whereby he became renued in spirit, body and excellency, De bubulio factus est Bajulus Regis Siciliae, from a Plow-man he was made Porter to the King of Sicily. And the Popes Letters assures us, That Almannus, held Prisoner by the Sar•cens, through the use of a Medicine lived five hundred years. For the King, whose Captive he

---

[20] Lib.22.cap.24.

was, having received this Medicine from the Embassadours of the great King, and being suspitious of them, made tryal hereof upon this Captive, which was brought him for that purpose. And the Lady of the Woods in great Britanny searching for a white Hinde, found an Ointment, wherewith the Keeper of the Woods anointed his whole body, except the soals of his feet, and he lived three hundred years without any corruption, save in the soals of his feet, which had some passions. We our selves know it frequent in these dayes, That plain Countrey men, without the advantage so much as of a Physicians advice, live very healthfully an hundred years, or little lesse. And these are the rather confirmed by the operations of Animals, as Harts, Eagles, Serpents, and many others, who by the efficacy of heart or stones, have renewed their youth: And wise men seeing, that even bruits could reach so farre to their Prolongation, adjudging it no lesse feasable by reasonable men, set themselves on the Spurre to find out this secret. Hereupon Artefius from his own ingenuity, having found the Secrets, of Stones, Herbs, Sensibles, &c. both for the knowledge of Nature, and especially the Prolongation of Life, did rejoyce,[21] that he had lived 1025. yeares. Further, to confirme this Assertion of the Prolongation of Life, it's considerable, That man naturally is

---

[21] A!. Glory

immortal, that is to say, Potens non mori, hath a possibility of not dying. Yea, even after his fall, he might live a thousand years, though by degrees the length of life was abbreviated. Hence it follows, That this abbreviation is Accidental, and consequentially may be repaired in whole or in part; and upon search we shall find the accidental cause of this corruption, is not from the Heavens, or any other than the defect of true Government of our health. In that our Fathers are corrupt and imbecil, they beget sonnes of a corrupt complexion and composition, and their children upon the same score are corrupted. Thus the Pedigree of corruption is deprived from Fathers to sonnes, untill we settle upon our heirs an assured abbreviation of our dayes. Yet this doth not conclude, That to perpetuity there shall succeed an abbreviation of our life, since there is a positive period set to our life, men may live till they be eighty years,[22] though then their dayes be but labour and sorrow.

Now if every man would from the brest exercise a compleat Regiment of health (which consists in such things as have relation to Meat, Drink, Sleep, Waking, Motion, Rest, Evacuation, Retention, Air, and the Passions of the mind.) He might find a remedy resisting his proper malady. For upon the

---

[22] Psalm 90.

prosecution of such a Regiment, one might arrive at the uttermost limit of that Nature he had from his Parents will permit,[23] and be led to the very last period of Nature (I mean Nature fallen from its original uprightnesse) beyond which there is no further progresse; because it doth little or nothing availe against the corruption of our Ancestours: and yet the great impossibility of any mans so ordering himself in a mean, in all the fore-mentioned things, as the Regiment of health exacts, wherefore abbreviation of our dayes does not only from our Progenitors, but hath its advantages from the want of Regiment. However the Art of Physick sufficiently determines this. Although nor rich, or poor, wise or ignorant, no nor the most accurate Physitians themselves, do accomplish this Regiment in themselves or others, as every eye can discern. Yet Nature is not deficient in Necessaries, or Art any wayes incompleat, but rather is advantagious to make insurrections and irruptions against, and so farre into these accidental passions, as they are either wholly or in part rooted out. At first, and in the beginning of our ages declining, the remedy was easie: But since we have five thousand years or more disadvantage, the Cure is more craggy.

---

[23] Deest parenthesis in alio exemplo.

But waving the Inconveniences wise men moved by the considerations forementioned, have endeavoured to find out the means and wayes, which not only are forceable against the defects of every mans proper Regiment, but also against the corruptions of our Parents: Not that hereby they can attain to the years of Adam or Artefius, by reason of the growing corruption, but that our dayes may be augmented an hundred yeares, or more, above the ordinary age of most men in these dayes. And though it be impossible absolutely to retard the accidents of old age, yet hereby they may mitigate them, so as life will happily be prorogued beyond the common account, yet alwayes within the ultimate circuit of Nature. There is a bounder of Nature, set in men since their Fall. There is a bounder of every particular man arising from the proper corruption of his Parents. Beyond both these bounders it's impossible to passe; yet happily one may arrive beyond the latter: nor yet so farre to go beyond it, as that the wisest of men can ever reach the former. Although there be a possibility and aptitude of Nature to proceed to that boundary our first Parents set them. Let no man think this strange, since this aptitude extends it self to immortality, as appears both before the fall, and shall be evident after the Resurrection.

Perhaps you may object, That neither Aristotle, Plato, Hippocrates, or Galen ever attained that prolongation. I shall answer, They have not attained the knowledge of many ordinary truths, which other ingenious heads have found out; and if so, they may easily miscarry in a businesse of such weighty consequence, though they made it their study: especially, if we consider, how they were burdened with other impertinencies, and so were sooner brought to their gray haires, spending the inch of their Candles in more debased and vulgar subjects, than in finding out the wayes to so great Secrets. We are not ignorant Aristotle sayes in his Predicaments, That the Quadrature of a Circle is possible, yet not then known. Yea he confesseth, himself and all his Predecessors were ignorant hereof, yet we in our times know it. Now if Aristotle did come short in such a trivial, much more might he in the deep mysteries of Nature.

Even in these dayes wise men are ignorant of many things, which the most ordinary capacity shall understand ere long. Thus the Objection is of little force.

Chap. VIII.

Of obscuring the Mysteries of Art and Nature.

After an enumeration of some few examples concerning the prevalency of Nature and Art (that by these few we may gather many,) by these parts the whole; and so from particulars, universals, which will demonstrate the unnessary aspiring to Magick, since both Nature and Art afford such sufficiencies. I shall now endeavour a methodical procedure in singulars, laying open both the causes and wayes in particular: and yet I will call to mind how as secrets (of [24] Nature) are not committed to Goats-skins and Sheeps-pelts, that every clown may understand them, if we follow Socrates or Aristotle. For the latter in his Secreta Secretarum affirmes, He breaketh the heavenly Seal, who communicateth the Secrets of Nature and Art; the disclosing of Secrets and Mysteries, producing many inconveniencies. In this case Aulus Gellius in Noct. Attic. de Collatione Sapientum, sayes, It's but folly to profer Lettices to an Asse, since hee's content with his Thistles. Et in lib. lapidum, The divulging of Mysteries is the diminution of their Majesty, nor indeed continues that to be a Secret, of which the whole fry of men is conscious.

---

[24] Derst in alie.

For that which all men, which wise, and the more
noted men affirme is truth. That therefore which is
held by the multitude, as a multitude, must be
false; I mean of that multitude, which is distinct
from knowing men. The multitude, it's true, agree
with wise men in the more vulgar conceptions of
their mind; but when they ascend to the proper
principles and conclusions of Sciences and Arts,
they much dissent (striving to get onely the
appearances in Sophismes and subtilties which wise
men altogether reject[25]). And this their ignorance of
the proprieties and Secrets, makes the division from
knowing men. Though the common conception of the
mind, have all one Rule and Agreement with knowing
men. Yet as for common things, they are of small
value, nor enquirable for themselves, but rather for
particular and proper ends.

The Reason then, why wise men have obscured their
Mysteries from the multitue, was, because of their
deriding and slighting wise mens Secrets of wisdome,
being also ignorant to make a right use of such
excellent matters. For if an accident help them to
the knowledge of a worthy Mystery, they wrest and
abuse it to the manifold inconvenience of persons
and communities. Hee's then not discreet, who writes
any Secret, unlesse he conceal it from the vulgar,

[25] Al. Vacans sophinatibus & inutilibus.

and make the more intelligent pay some labour and sweat before they understand it. In this stream the whole fleet of wise men have sailed from the beginning of all, obscuring many wayes the abstruser parts of wisdome from the capacity of the generality. Some by Characters and Verses have delivered many Secrets. Others by aenigmatical and figurative words, as Aristotle sayes, in lib. Secret, O Alexander, I shall disclose to you the greatest of Secrets, which it becomes you by divine Assistance to keep secret, and perfect the thing proposed. Take then then the Stone, which is no Stone, which is in every man, and in every place, and in all times; and it shall be called the Philosophers Egge, and the Terminus Ovi. And thus we find multitudes of things obscured in the Writings and Sciences of men, which no man without his Teacher can unvail.

Thirdly, They have obscured their Secrets by their manner of Writing, as by Consonants without Vowels, none knowing how to read them, unlesse he know the signification of those words.[26] Thus the Hebrewes, Caldees, Arabians, nay the major part of men do most an end write their Secrets, which causeth a great obscurity amongst them, especially amongst the Hebrewes. For, as Aristotle sayes in his fore

_____

[26] Significata/

246

recited Book, God gave them all manner of wisdom long before they were Philosophers: And all Nations had their Originals of Philosophy from the Hebrewes, as Albumazar in lib. Introductorii Majoris; and other Philosophers, with Iosephus lib 1. & lib. 8. Antiquit. makes it evident.

Fourthly, This obscuring is occasioned by the mixture of several sorts of Letters,[27] for so the Eth•ick Astronomer hid his knowledge, writing it in Hebrew, Greek and Latine Letters altogether.

Fifthly, This obscuring was by their inventing other letters, then those which were in use in their own, or any other Nation, being framed meerly by the pattern of their own fancy, which surely is the greatest impediment; yet this was the practice of Artefius in lib. de Secretis Naturae.

Sixthly, They used not the Characters of Letters, but other Geometrical Characters, which have the power of Letters according to the several Position of Points, and Markes. And these he likewise made use of.

---

[27] Ethicus Astronomu, fortasse N. acest ergo Anglice dedi Ethnick.

Seventhly, There is a greater Art of obscuring, which is called Ars Notoria, which is the Art of Noting and Writing, with what brevity, and in what manner we desire. This way the Latines have delivered many things. I held it necessary to touch at these obscurings, because it may fall out, I shall thorow the magnitude of our Secrets discourse this way, that so I may help you so farre as I may.

Chap. IX.

In aliis Adverg.

Of the Manner to make the Philosophers Egge.

Now I shall methodically handle those things I promised above,[28] the dissolving the Philosophers Egge, and finding out the parts thereof; a work which will give beginning to other enterprises. Make a diligent purification of the Calx with the waters of Alkali, and other acute waters, grind it by several contrition with the salts, and[29] burn it with many assations, that the earth may be perfectly separated from other elements, which I hold worthy the[30] longitude of my stature. Understand it if you can. For without doubt there will be a composition of Elements, and so it will be part of that Stone which is no Stone, which is in every man, and in every place of man; and you may find this in all the seasons of the year in its place. Then take oyl after the form of a Saffron-cheese,[31] and so viscouous first (as not to be smitten asunder by a stroak) divide the whole fiery virtue, and separate it by dissolution, and let it be dissolved in acute water, of a temperate acutenesse, with a slight

---

[28] These are aenigmatical.
[29] Al. Contermina.
[30] Al. Melancholia staturae.
[31] Al. Insensibile.

fire, and let it be boyled till his[32] fatnesse, as the fatnesse of flesh be separated by distillation, that nothing of the unctiousnesse may issue forth; and let this fiery virtue be distilled in the water of Urine. Afterwards boil it in Vinegar, untill the least part, which is the cause of adustion be dried up, and the fiery virtue may be had; but if theere be no regard of it,[33] again let it be made. Mind and search what I say: for the speech is difficult. The Oyl is dissolved in acute waters, or in common Oil which works more expressly,[34] or in acute Oyl of Almonds upon the fire; so as the Oyl be separated, and the spirit remain occult, in the parts of living creatures, Sulphur and Arsnick. For the stones, in which the Oyl of humidity overflows, have their terminus in the union of its parts: for there is no vehement union, but one may be dissolved from another by the nature of water, which is the subject of liquefaction in the spirit, which is the Medium betwixt the dry parts and the Oyl. The dissolution being made there will remain in the spirit, a pure humidity, vehemently mixed with dry parts, which are moved in it, when the fire resolves it, which is sometimes called of the Philosophers, Sulphur fusibile, sometimes Oyl, other while an aery humour, sometime a conjunctive substance, which the fire

---

[32] Al. Terrestreitas.
[33] Al. Tumfac.
[34] Al. Vt.

separates not, sometimes Camphore: and if you
please, this is the Philosophers Egge, or rather the
Terminus and end of the Egge; and it came to us from
these Oyls, and may be esteemed amongst the
subtilties, when it is purged and separated from the
water and oyl in which it is. Further, the Oyl is
corrupted by grinding it with desiccating things, as
with salt or Atrimentum, and by assation, because
there is a passion arising from the contrary; and
afterwards it is to be sublimated, untill it be
deprived of[35] its oleagmeity, and because its as
Sulphur or Arsnick amongst Minerals, it may be
prepared, even as it. Yet it's better to boil it in
waters, that are temperate in acuity, untill it be
purged and whitened. Which wholsom exaltation is
made either in hot or moist fire: The distillation
must be re-iterated, that it may sufficiently
receive its goodnesse, untill it be rectified, the
signs of its last rectification are candor and
crystalline serenity: And when other things grow
black by fire, this grows white, is cleansed, shines
with clearnesse and admirable splendour. From this
water and its earth comes Argentum vivum in
Minerals, and[36] when the matter hath waxed white,
this way it is congealed; the Stone of Aristotle,
which is no Stone, it's set in a Pyramid a hot

---

[35] Al. Olio suo.
[36] Al. Quandoque in salem Ardoniacum.

place, or (if you please) in the belly of an Horse or Ox, and it imitateth an acute Feavor. For from seven to fourteen, and from that it sometimes proceeds to one and twenty, that the Fecis of the Elements may be dissolved in its water, before it be separate: The dissolution and distillation is to be iterated, untill it be rectified. And here is the end of this intention. Yet know that when you have consummated your work, you are then to begin.

Another Secret I shall shew you, you must prepare Argentum vivum by mortifying it with the vapour of Tin for Pearls, and with the vapour of Lead for the Stone Iberus; then let it be ground with desiccating things, and Attramentis, and the like, as is said, and let there be an assation: Then let there be a sublimation[37] if for Pearles twelve times; if for rednesse one and twenty times, untill the humidity within it be totally corrupted. Nor is it possible, that its humidity be separated by vapour, as the fore-said oyl; because its vehemently mixed with its ary parts; nor doth it constitute, as in the foresaid metals. In this Chapter you may be deceived, unlesse you distinguish of the signification of the words. It's now high time I involve the third Chapter: that you acquire the Calx, the[38] Calx of the body, which you intend, the

---

[37] Al. septies.

252

body is calcined, when it is appodiated, i.e. that
the humour in it may be corrupted by salt, and with
salt Armoniack and vinegar, and sometimes with
burning things, and with Sulphur and Arsnick: and
sometimes bodies are fed with Argentum vivum, and
sublimated from them, untill they remain[39] putred.
The Claves of the Art are congelation, resolution,
inceration, proportion; and another way
purification, distillation, seperation, calcination
and fixation, and then you may acquiesce.

---

[38] Al. Clavem operis.
[39] Remaneant putris, ni fallor male imprimitar aut debet esse
pura aut putres.

Chap. X.

Forte. 620.

Of the same subject another way.

In the 602. Arabian year you intreated me for some Secrets. Take then the Stone, and calcine it with a light assation and strong contrition, or with acute things. But in the end mingle it a little with sweet water, and compound a Laxative Medicine of seven things (if you will) or of six, or of five, or as many as you please; but my mind rests in two things, whose proportion is better than the other sixt proportion, or thereabouts, as experience will teach you.[40] Resolve notwithstanding the gold at the fire, and tried it better; but if you will credit me, take one thing that is the Secret of Secrets of Nature, able to do Miracles. Let it be mixed from two or more, or a Phoenix, which is a singular creature † at the fire, and incorporate by a strong motion: to which if hot liquor four or five times be applied, you have the composition. Yet afterwards the coelestial nature is debilitated, if you infuse hot water three or four times. Divide therefore the weak from the strong in several vessels, if you believe me: Let that which is good be evacuated. Again, use the pouder, and the water which remains, carefully expresse: For of a certain, it will produce the

---

[40] Al. Mollius calescent.

parts of the pouder, not incorporated; therefore take the water by it self; because the pouder exiccated from it hath power to be incorporated into the Laxative Medicine. Work therefore as formerly, untill you distinguish the strong from the weak, and apply the pouder three, four, five times or oftener, and work alwayes the way: And if you cannot work with hot waters, do it with water of Alkali, and by such acute things you make the violence of the Medicine. But if by reason of the acuity and softnesse of the Medicine it be broken, the pouder, being applied, apply very carefully more of the hard and soft. But if it be by reason of the abundance of the pouder apply more of the Medicine; if it be by reason of the strength of the water, water it with pistils;[41] and congregate the matter, as you can, and separate the water by little and little, and it will return to its state, which water you must exiccate: for it contains both pouder and water of the Medicine, which are to be incorporated, as the principle pouder. Here you may not sleep, because here is contained a very great and profitable Secret. If you rightly order in a right series of things, the parts of the Shrub or Willow, they will keep natural union: and do not deliver this to oblivion, for it is profitable for many things.[42] You

---

[41] Pistillo.
[42] Unione facta.

must mingle Pearls with the made union: as I think there will arise something like the Stone Iberus: and without doubt it mortifies that which is to be mortified by the vapour of Lead. You shall find Lead, if you expresse the living from the dead; and the dead you must bury in Olibanum and Sarcocolla. Keep this Secret, for it is of some profit, and so must you do with the vapour of Pearls, and the Stone Tagus, and you must (as I have said) bury the dead.

Chap. XI.

Forte. 603.

Of the same Subject another way.

To your desire in the Arabian year 630. I return
this Answer. You must have the Medicine which may be
dissolved in the thing liquified and steeped in it,
and penetrate its interior parts, and may be mingled
with it; and it may not be a fugitive servant, but
transmute it. Let it be mingled by reason of the
spirit, and let it be fixed by the Calx of the
metal: it is to be thought that fixion is prepared,
when the body and spirit are set in its place, and
the spirit is made a body. Take then of the bones of
Adam, and of the Calx the same weight; let there be
six to the Stone Tagi, and five to the stone of
Pearl; let them be ground with Aqua vitae, whose
property it is to dissolve all other things, so as
in it they are dissolved and assated, untill it be
incerated, i.e. let the parts be united, as the
parts in wax. The sign of inceration is, that the
Medicine liquifies upon iron very hot. Then let it
be put in the same water in some hot and moist
place, or let it hang in the vapour of waters made
very hot: after that dissolve and congeal them
against the Sunne. Afterwards take Saltpeter, and
argentum vivum shall be converted into lead: And
again, wash the lead with it, and mundifie it, that

it may be the next to silver, and then work as a pious man, and also the whole weight must be 30. But yet of Saltpeter LVRVVOPO Vir Can Vtriet Sulphuris: and so you may make Thunder and Lightning, if you understand the Artifice: but you must observe, whether I speak aenigmatically, or according to the truth. Some men have supposed otherwise: For it is told me, that you must resolve all into its first matter, of which you have Aristotle speaking in vulgar and known places, which makes me silent herein. When you have this, you have pure, simple and equal Elements. And this you may do by contrary thing and various operations, which formerly I have called the Claves of the Art. And Aristotle sayes, That the equality of potencies excludes action, and passion, and corruption. And these things Averrho's affirms, reproving Galen. And this Medicine is esteemed the more pure and simple which may be found, which is prevalent against Feavers, passions of the mind and body. Farewell. Whoever unlocks these, hath a key which opens and no man shuts: and when he hath shut no man opens.

In this Translation, I followed Dr. Dees Edition, Printed at Hamburg, 1618.

FINIS.

A Word from the Publisher

Thank you for purchasing this small work from The R.A.M.S. Library of Alchemy. During his lifetime, Hans Nintzel was dedicated to the identification, acquisition, study, retyping and, when necessary, translation of what he considered to be the most important known works on Alchemy. Hans was assisted by his sparse network of fellow Alchemists, all members of the Restorers of Alchemical Manuscripts Society (R.A.M.S.). I was an active member of R.A.M.S.

My goal is to publish all of the works originally made available through R.A.M.S. as photocopies. To facilitate this, I have chosen to have the books professionally printed. I also have a few titles that I intend to add to the original R.A.M.S. Library, selected by strict criteria established by Hans.

The works from the original R.A.M.S. Library are republished by R.A.M.S. Publishing Company in the collection, "The R.A.M.S. Library of Alchemy," with permission of the Estate of Hans W. Nintzel.

If you have a work on Alchemy that you believe should be a part of the R.A.M.S. Library, please contact me through R.A.M.S. Publishing Company.

Philip N. Wheeler

www.ingramcontent.com/pod-product-compliance
Lightning Source LLC
Chambersburg PA
CBHW080802180526
45168CB00006B/2299